National Academy Press

The National Academy Press was created by the National Academy of
Sciences to publish the reports issued by the Academy and by the
National Academy of Engineering, the Institute of Medicine, and the
National Research Council, all operating under the charter granted to
the National Academy of Sciences by the Congress of the United States.

Safety of Dams

Flood and Earthquake Criteria

Committee on Safety Criteria for Dams
Water Science and Technology Board
Commission on Engineering and Technical Systems
National Research Council

NATIONAL ACADEMY PRESS
Washington, D.C. 1985

NATIONAL ACADEMY PRESS   2101 Constitution Ave., NW   Washington, DC 20418

NOTICE: The project that is the subject of this report was approved by the Governing Board of the National Research Council, whose members are drawn from the councils of the National Academy of Sciences, the National Academy of Engineering, and the Institute of Medicine. The members of the committee responsible for the report were chosen for their special competences and with regard for appropriate balance.

   This report has been reviewed by a group other than the authors, according to procedures approved by a Report Review Committee consisting of members of the National Academy of Sciences, the National Academy of Engineering, and the Institute of Medicine.

   The National Research Council was established by the National Academy of Sciences in 1916 to associate the broad community of science and technology with the Academy's purposes of furthering knowledge and of advising the federal government. The Council operates in accordance with general policies determined by the Academy under the authority of its congressional charter of 1863, which establishes the Academy as a private, nonprofit, self-governing membership corporation. The Council has become the principal operating agency of both the National Academy of Sciences and the National Academy of Engineering in the conduct of their services to the government, the public, and the scientific and engineering communities. It is administered jointly by both Academies and the Institute of Medicine. The National Academy of Engineering and the Institute of Medicine were established in 1964 and 1970, respectively, under the charter of the National Academy of Sciences.

   This report represents work supported by Cooperative Agreement Number 4-FC-81-04270 between the Bureau of Reclamation (contracting agency for the Corps of Engineers and Bureau of Reclamation) and the National Academy of Sciences.

Library of Congress Cataloging in Publication Data
Main entry under title:

Safety of dams.

   Bibliography: p.
   Includes index.
   1. Dam safety—United States.   2. Flood control—United
States.   3. Earthquakes—United States.   I. National
Research Council (U.S.).   Committee on Safety Criteria for
Dams.
TC556.S24   1984   363.3'497   85-2973
ISBN 0-309-03532-5

COMMITTEE ON SAFETY CRITERIA FOR DAMS

GEORGE W. HOUSNER, *Chairman*, California Institute of Technology, Pasadena
KEIITI AKI, University of Southern California, Los Angeles
DONALD H. BABBITT, California Division of Safety Dams, Sacramento
DENIS BINDER, Western New England College, Springfield, Massachusetts
CATALINO B. CECILIO, Pacific Gas and Electric Company, San Francisco, California
ALLEN T. CHWANG, The University of Iowa, Iowa City
MERLIN D. COPEN, Consultant, Aurora, Colorado
LLEWELLYN L. CROSS, Charles T. Main, Inc., Boston, Massachusetts
CHARLES H. GARDNER, North Carolina Department of Natural Resources, Raleigh
LESTER B. LAVE, Carnegie-Mellon University, Pittsburgh, Pennsylvania
DOUGLAS E. MACLEAN, University of Maryland, College Park
OTTO W. NUTTLI, St. Louis University, St. Louis, Missouri
JOHN T. RIEDEL, Hydrometeorological Consultant, Huddleston, Virginia
GURMUKH S. SARKARIA, International Engineering, Co., Inc.
H. BOLTON SEED, University of California, Berkeley
JERY R. STEDINGER, Cornell University, Ithaca, New York

Ex-Officio Members

WALTER R. LYNN, Cornell University, Ithaca, New York; Chairman, Water Science and Technology Board

ROBERT L. SMITH, University of Kansas, Lawrence; Member, Commission on Engineering and Technical Systems; Member, Water Science and Technology Board

iii

Technical Consultant

HOMER B. WILLIS, Consulting Engineer, Bethesda, Maryland

NRC Project Manager

STEPHEN D. PARKER, Executive Director, Water Science and Technology Board

NRC Project Secretary

JEANNE AQUILINO

Federal Agency Technical Representatives

DONALD DUNCAN, U.S. Army Corps of Engineers, Washington, D.C.
DAVID PROSSER, U.S. Bureau of Reclamation, Washington, D.C.

v

Preface

Many thousands of dams have been constructed in the United States, and new dams continue to add to this total. The proper functioning of these dams under all conditions is an important matter of public safety and welfare. This report concerns the levels of safety to be provided at dams to withstand extreme floods and earthquakes.

The occasional failure of a dam stimulates public concern, and in response, safety assessments are undertaken. Such assessments have recently been made by the U.S. Army Corps of Engineers and the Bureau of Reclamation, two federal agencies having major dam programs. However, many dams are also constructed and operated by other federal, state, and local government agencies, utilities, corporations, and individual owners. The study for this report was undertaken at the request of two departments of the federal government, but the report is for the most part relevant to all dams, both federal and nonfederal.

On the average, about 10 significant dam failures have occurred somewhere in the world in each decade, and many more damaging near-failures have occurred. Some of these events have resulted from incorrect decisions made during the design and construction process, whereas others have been the consequence of inadequate maintenance or operational mismanagement. Many have resulted from unanticipated large floods, and a few have resulted from intense earthquake shaking. The water retained in a large reservoir has enormous potential energy that can cause extensive loss of life and damage to property. In fact, few activities of man pose greater potential for destruction. Accordingly, engineers tend to take a very conservative approach in designing dams; however, the more conservative the design, the greater the cost of safety. Also, relatively few dams will experience the extreme events for which they are designed, but the location and magnitude of these events cannot be predicted and, therefore, conservative designs generally are provided at most dams to avoid catastrophic failures at a few.

Earthquakes and floods pose a similar problem to designers of dams, in that both hazards have uncertainties associated with the occurrences of extreme events and decisions must be made as to the best way to handle these uncertainties. However, there are also significant differences in the problems posed. For example, since most dams are built to retain runoff from a watershed, questions of extreme floods usually arise. On the other hand, some regions of the United States have a low seismic hazard, and ground shaking does not pose a serious threat to the safety of dams.

The Committee on Safety Criteria for Dams was requested to report on the selection of appropriate flood and earthquake occurrences to be considered in design of dams and safety evaluation of dams. This report represents a general consensus of the views and conclusions of the committee. Although the committee did not attempt to make the report a treatise on protecting dams from earthquakes and extreme floods, it did include background material to aid in understanding the bases for its findings and recommendations. The time and funding available to the committee precluded the undertaking of in-depth studies and research to develop detailed, new design criteria. Rather, the committee (1) reviewed current practices in the United States and abroad in regard to designing dams for extreme hydrologic and seismic events and (2) made recommendations for action and research aimed at improving safety evaluations of dams with respect to extreme events. The committee found that its deliberations led to questions of risk and responsibility; therefore, the report also addresses these matters.

The members of this committee recognize that they have participated in an important and unusual activity, and they appreciate the responsibility this assignment has placed upon them. In most instances the formulation, or the review, of criteria for engineering work is accomplished by a group of similarly minded specialists in a narrow branch of technology. It would be difficult to find a parallel to the assignment and composition of this committee. The charge to the group was a difficult one, and there are issues on which the committee did not reach complete agreement and the report recommendations represent a consensus of views. One particular issue of concern to the committee is the continued use of the probable maximum flood (PMF) as a principal basis for design of spillways for all new dams in high-hazard situations. Some committee members felt such a design basis, in some cases, results in extravagant use of resources, but they also recognized that an adequate substitute design basis is not available at this time. Another concern involves the lack of quantitative definitions for the dam hazard classifications used by federal and state agencies along with a lack of uniformity in the spillway design floods assigned to each hazard and size category. The committee noted a considerable variety in these standards. Some members proposed that the committee attempt to formulate "hazard classification" standards that could be recommended; however, time did not permit such an effort, and the consensus was that such an activity should be pursued by other groups.

The committee concluded, on the basis of information presented, that the Corps of Engineers, the Bureau of Reclamation, and some other federal agencies, as well as some state agencies and engineering firms, are generally using up-to-date methods of assessing flood and earthquake hazards. The committee felt that up-to-date methods could be further improved by research and by collection of relevant data.

The importance to the nation of the problems of the safety of dams against extreme floods and earthquakes is widely recognized, and the committee foresees that the importance of these problems will increase as population density increases and water becomes a scarce resource. Thus, protection of dams against such events should receive the continued attention of federal and state governments, as well as the relevant engineering and science communities.

The committee has been aided greatly in its work by many people and organizations. In the Acknowledgments that follow, some of the contributions to this effort are briefly mentioned. For the committee, I express gratitude for this help. For myself, I wish to thank all the committee members, members of the Water Science and Technology Board, members of the National Research Council and federal agency staffs, the technical consultant, and others who have inspired and facilitated the task at hand.

GEORGE W. HOUSNER, *Chairman*
Committee on Safety Criteria for Dams

Acknowledgments

Although the committee takes full responsibility for the material in this report, it wishes to acknowledge valuable contributions to the committee's work from many organizations and individuals, including the following: those organizations who responded to the requests for information on dam safety criteria in use relating to extreme floods and earthquakes; staff members of a number of federal agencies, including the Bureau of Reclamation, U.S. Army Corps of Engineers, National Weather Service, U.S. Geological Survey, Tennessee Valley Authority, Soil Conservation Service, Forest Service, Federal Energy Regulatory Commission, Federal Emergency Management Agency, and Nuclear Regulatory Commission, who contributed to the committee's discussion and understanding of current practices and problems; and the employers of the members of the committee who generously made those members available for the committee sessions.

Contents

Appendixes

List of Major
Figures and Tables

Figures

xv

List of Major Tables and Figures

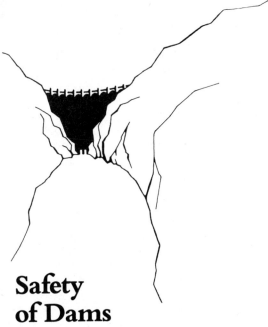

Safety
of Dams
Flood and Earthquake Criteria

Executive Summary

In response to a request by the Assistant Secretary of the Interior for Water and Science and the Assistant Secretary of the Army for Civil Works, the National Research Council established an ad hoc Committee on Safety Criteria for Dams to prepare an inventory of currently used criteria for dams relating to safety from hazards of extreme floods and earthquakes, and to identify and evaluate alternative criteria for safety of federal dams. The committee was made up of recognized experts in risk assessment, regulation of dams in the interest of public safety, law, and science and engineering, most of them involved with consideration of safety of dams from hazards of extreme floods and earthquakes. The committee was assisted in its analysis of the problems of protecting dams and those persons and properties downstream as well as upstream from dams from flood and earthquake hazards by several federal agency representatives and a great many organizations and individuals who provided information on current practices. The committee considered a wide variety of issues ranging from the sciences of meteorology and earth tectonics to the role of government and principles of common law. This report presents the results of the committee's work and its recommendations. The prime focus of the committee's considerations was the protection of dams from failures caused by floods and earthquakes and, thereby, the protection of people and properties from the hazards of dam failures. The committee's principal findings and conclusions are summarized as follows:

1. Extreme floods and earthquakes are random events that can but may not occur during the life of a dam.

1

2. Current dam design practices reflect great diversity in the standards used for classifying dams relative to hazards and in the criteria for evaluation of safety from extreme floods and earthquakes. Of the organizations responding to the committee's request for information, the U.S. Army Corps of Engineers, the Bureau of Reclamation, and some others apply state-of-the-art methods in evaluating safety of dams from floods and earthquakes.

3. While simple hazard rating categories based on downstream development may be useful for identifying dams for high-priority safety evaluation and study, they do not reflect the potential for incremental loss of life and damage caused by the failure of a dam due to an inadequate spillway when a river is already in flood.

4. More uniformity is needed among the several federal and state agencies establishing size and hazard definitions and correlative design standards. (The committee has not recommended size and hazard definitions in this report, but reference to hazard classifications in the report correspond in a general way to those used in the National Program of Inspection of Non-Federal Dams conducted by the U.S. Army Corps of Engineers.)

5. Many of the basic concepts for estimating extreme flood-producing capabilities of watersheds have been used for several decades. However, new concepts and improved methods for estimating floods, changes in patterns of land use and an expanding data base have resulted in generally larger flood estimates. For these same reasons, in general future estimates of magnitudes of extreme floods can be expected to continue to increase. However, unless the runoff characteristic of the watershed were to change, increments in future estimates for a specific basin resulting from extension of the data base should, in general, be less than those noted in the past. Also, it is noted that there have been instances of later, more intensive hydrometeorological studies that resulted in reductions in estimates of probable maximum precipitation potentials from those established by earlier investigations.

6. A dam designed for the estimated probable maximum flood, based on the estimated probable maximum precipitation as determined by current practices, does not necessarily provide absolute assurance that the dam is safe for every possible flood.

7. During the past 20 years, there have been major advances in our knowledge of earthquake ground motions, in our abilities to predict resultant structural response, and in the methods of analyzing the safety of dams against earthquakes.

8. Methods of estimating potential earthquake ground motions are becoming more reliable in regions with well-identified fault systems (such as the western United States) and also in regions where fault systems are not well identified (such as the eastern United States).

9. Currently, three basic approaches are used to determine the magni-

tude of floods for which the safety of dams should be evaluated (i.e., the deterministic, probabilistic, and risk analysis approaches). Each approach has advantages and disadvantages that must be evaluated in selecting an appropriate method for any specific project.

10. Currently, three basic approaches are used to determine the magnitude of earthquake motions for which a dam will be evaluated (i.e., the deterministic-statistical, seismotectonic province, and probabilistic-risk analysis approaches). At the present time, the deterministic-statistical and the seismotectonic approaches are the most widely used.

11. Risks are present in every human activity. The federal government has become increasingly involved in risk management issues in the last two decades. A number of federal agencies have developed various risk management standards for specific types of risk, but these are not applicable to risk management for dams. Adaptation of these concepts to dam safety requires research that has not been done by this committee.

12. Court decisions relating to dam failures in general have held the owner liable for the damages resulting from a failure.

The committee recommends the following:

1. To the extent practicable, reservoir safety evaluations should strike a balance among such considerations as project benefits, construction costs, social costs, and public safety, including the possible consequences of dam failure due to major earthquakes and floods. (While achieving such balance is the ideal, currently available technology does not permit this balancing with full confidence in the results.)

2. Safety evaluation standards for *existing* dams and *proposed* dams need not necessarily be the same.

3. The use of probable maximum floods (PMF), based on estimated probable maximum precipitation (PMP), as the general design standard (safety evaluation flood (SEF)) for proposed high-hazard dams should be continued. However, instances sometimes will be encountered where a lower standard may be justified if failure of a dam during floods of PMF magnitude would cause no significant increase in potential for loss of life or property damage.

4. For existing high-hazard dams, the adopted safety evaluation flood (SEF) should take into account estimated flood probabilities, expected project performance, and incremental damages that would result from dam failure for a range of floods up to and including the probable maximum floods.

5. In the design of new dams and spillways when design alternatives of approximately equal cost are available, a selection among these alternatives

should give consideration to potential future needs for increased safety against extreme floods and earthquakes.

6. The maximum credible earthquake (MCE) ground motions developed by the deterministic-statistical method applied to known causative faults provide an acceptable level of conservatism for safety analyses of high-hazard dams; this approach should be used whenever possible. However, research should be continued to develop probabilistic-risk analysis methods. Where earthquake sources are not well identified, the seismotectonic province method for determining MCE should be used.

7. The maximum credible earthquake (MCE) should be adopted as the safety evaluation earthquake (SEE) for a high-hazard dam; but the SEE for a lower-hazard dam may, in some cases, be less severe.

8. Response of a high-hazard dam to earthquakes in regions of significant seismicity should be analyzed utilizing dynamic analysis techniques, combined with appropriate judgment, with the objective of assuring that the safety evaluation earthquake will not cause catastrophic release of water from a reservoir.

9. The safety evaluation of dams need not consider the simultaneous occurrence of the safety evaluation flood and the safety evaluation earthquake because of the extremely low probability of such occurrence.

10. Periodic reviews of hazard determinations and safety decisions for all dams should be required, especially when safety evaluations are based on criteria less conservative than the probable maximum flood or the maximum credible earthquake.

11. Research efforts designed to provide better bases for estimating magnitudes and frequencies of extreme floods and earthquakes, for estimating reactions of dams to such natural phenomena, and for establishing acceptable levels of risks should be continued.

12. As advances occur in seismology, hydrology, meteorology, and the relevant data bases, and as changes are noted in public attitudes toward risk, the federal agencies should periodically undertake a review of dam safety practices and standards by an independent body representing the professions involved in engineering for dams and experts from other relevant disciplines.

1
Introduction

In May 1984, subsequent to a request from the Assistant Secretary of the Interior for Water and Science and the Assistant Secretary of the Army for Civil Works, the National Research Council initiated a study of criteria for the evaluation of the adequacy of spillways and earthquake resistance of dams. An enabling cooperative agreement provided that the Council would establish a Committee on Safety Criteria for Dams; prepare an inventory of existing safety criteria in use by federal, state, and nongovernmental entities in the design of dams, particularly with respect to spillway capacities and earthquake hazard, to serve as an information base for consideration by the committee; and identify and evaluate alternative criteria for establishing minimum levels of safety for federally designed, constructed, or operated dams. The committee was to consider both deterministic and frequency-based criteria, as applied to extreme flood and earthquake events. In identifying alternative criteria it was considered essential to reflect the influences of various natural and man-made conditions, size of dam, and probable effects of dam failure. Further, it was agreed that new dams and existing dams should be considered separately. A comparison of the alternative criteria, including the impacts, costs, and other implications of their use, was agreed to be an important and integral part of the evaluation process.

A Committee on Safety Criteria for Dams was appointed to function under the auspices of the Water Science and Technology Board of National Research Council. The committee included experts in hydrology, hydraulics, general dam engineering, seismology, geology, meteorology, earthquake engineering, economics, law, and risk assessment. The Committee on

5

Safety Criteria for Dams, with a December 1984 completion deadline, arranged a brief but intensive study of its assigned task.

Simultaneous with the appointment of the committee, input to an inventory of existing safety-related criteria for dams was sought from essentially every type of organization that might have an interest in dam safety. This inventory was created in an effort to assure consideration by the committee of existing approaches to dam safety standards. In May 1984, two meetings of several committee members, the government technical representatives, and the Water Science and Technology Board members and staff were held to organize the study, sort out assignments, and prepare for working meetings of the full committee. The committee met with experts from several of the federal water resources agencies in Washington, D.C., on June 25-26 and August 13-14, 1984, to discuss, review, debate, and develop draft documents. A draft report was produced from the activities of these meetings, and the committee refined its report at a subsequent meeting in Pasadena, California, on September 20-21, 1984.

Throughout the study process, the committee was acutely aware of the importance of its report to those responsible for safety of the nation's several thousands of dams. Although the focus of the committee's considerations has been on the dams for which the federal government is primarily responsible, the potential effect of the committee's report on the dam safety programs of state governments has been kept in mind. It is recognized that the level of effort appropriate to evaluating the safety of each of the major federal dams may not be suitable to the differing circumstances encountered in state dam safety programs and in other nonfederal dam safety activities. Such aspects as limitations in available financial and technical resources, differences in size of dams involved, and differences in the general levels and extent of potential hazards may justify use of simplified procedures and criteria, particularly for preliminary screening of large numbers of dams in the interest of public safety.

The request by the Departments of Army and the Interior for the National Research Council to undertake this study is one more indication of the high level of interest in dam safety that has been evident throughout the world in the past few years. This interest has been brought about by such developments as the following:

• The occurrence of several disastrous incidents involving uncontrolled releases of impounded waters in the last two decades (e.g., Vaiont in Italy, Malpasset in France, and Machha II in India, and in the United States such dams as Buffalo Creek, Bear Wallow, Teton, Toccoa Falls, and Laurel Run) and several near failures (severe damage) from earthquakes (e.g., Koyna

Dam in India and the Upper and Lower Van Norman (San Fernando) Dams in California).

• The results of the National Dam Inspection Program conducted by the U.S. Army Corps of Engineers, which found that one-third of the approximately 9,000 dams inspected (all in high-hazard situations) were tentatively classified as unsafe.

• The realization that improvement of dams in the United States to meet current safety standards would have very high costs and that finding funds to improve many dams would be difficult.

• The increased activity in many states in regulating privately owned dams in the interest of public safety.

All of these developments contributed to the general perception that dam safety criteria are important to many interests. However, three other developments may be regarded as proximate causes for the request for this study:

1. The finding, based on investigation of existing projects by the Bureau of Reclamation and the Corps of Engineers in light of current concepts of the effects of earthquakes on dams, that some of those projects need alteration to assure acceptable safety during earthquakes.

2. Findings that many existing dams fail to meet current spillway capacity criteria for new dams.

3. Tightened budget requirements at the federal level and need for increased justification for requests for funds to improve existing federal dams in the interest of public safety.

The committee is gratified that, through its work, a considerable range of professional viewpoints and backgrounds has been brought to bear on the problems involved in dam safety criteria relating to extreme floods and earthquakes. However, establishing such criteria necessarily involves balancing risks among various interests. Establishing acceptable levels of risk to humans and properties is a matter of public policy. Technical experts can only help to determine what level of risks are acceptable. Guidelines for dam safety should not be applied in a mechanical fashion by specialists in a narrow field of technical activity. A review in the broad public interest of dam safety evaluations and plans to improve dam safety is desirable for each project.

2

Extreme Floods and Earthquakes—The Nature of the Problem

Other chapters of this report describe the technical methods that have evolved to estimate the magnitude of extreme floods and earthquakes, the limitations of the methods, and some possible improvements in the methods. In this chapter an attempt is made to take a broader look at the problems of coping with extreme floods and earthquakes at dams from the viewpoint of society as a whole. In so doing, it will be shown why there are no absolute solutions to these types of problems.

DESIGN OBJECTIVES

Dams are designed and constructed to withstand various natural forces and events that have occurred in the past or may be expected to occur in the future. A vital part of the process of designing these structures, which generally are expected to serve society for 100 years or more, is to anticipate the future vagaries of nature that may result in floods or earthquakes that would cause the dam to fail (i.e., be breached or collapse).

Some people have the mistaken impression that dams, especially government-built flood control dams, are designed and operated to protect property and residents downstream against all floods that could conceivably occur. This is not usually true. Extreme events could overwhelm the flood storage capacity of even large reservoirs. When such extreme floods occur, spillways pass on the large inflows, possibly leading to downstream flooding. To set this matter in perspective, it is important to understand that the primary purpose of a dam spillway is to *protect the dam* itself from failing

8

due to breaching or overtopping. A properly functioning spillway protects the dam by passing excess flood waters downstream, thereby limiting the amount of water held behind the dam. Thus an extraordinarily large flood may pass over the spillway and cause damages downstream possibly approaching those that would have occurred if the dam had not been built. However, the failure of the dam could produce flood rates and damages greater than would have been experienced if the dam had not been built.

While either exceptionally severe earthquakes or extremely large floods could cause a dam to fail, in this chapter the main attention is devoted to floods. The reason for this is simply that essentially *all* dams are exposed to the threat (and reality) of floods, whereas a much smaller fraction of existing dams may be subjected to significant earthquake forces (i.e., those located in active seismic zones). Also, the structural characteristics of many dams provide them with an inherent capacity to withstand earthquake forces, while protecting a dam against failure from overtopping can only be achieved in most cases by providing specific flood handling facilities (i.e., spillways).

Experience indicates that rare and large-magnitude precipitation events can produce flows of water with which most dam structures cannot cope—except to pass them downstream. When considering such floods, it would be very desirable to be able to predict how frequently extreme events of specific magnitude might be expected to occur. Some argue that only after developing a satisfactory response to this problem can one rationally determine the degree of protection from floods that should be provided in dam structures.

Surprising as it may seem to some, under criteria in current use to determine capacity built into dam structures to withstand or pass floods, it is possible for a flood event to exceed the dam's capacity to resist it. While terms such as unlikely, rare, or low probability are used to describe these kinds of events, it should be understood that what is being described is an event that can occur. *The problem faced by designers of dams and members of the body politic who use, pay for, or are affected by these structures is to determine "how much protection should be provided for the dam," considering these events can but may not occur during the life of the dam.* It is not possible to provide *absolute* safety against all hazards and especially from events produced by "mother nature." The objective should be to balance the benefits of making dams safer against the cost of the increased safety and to reduce any risks to acceptable proportions.

Objectives for either design or safety evaluations of dams relating to extreme floods and earthquakes can be considered in two broad categories, namely, (1) those relating to economic efficiency and (2) those relating to equity. The economic efficiency objectives encourage maximizing the excess of project benefits over project costs. Equity objectives seek appropriate balance between competing interests of such parties as the dam owner, those who benefit from the dam, and those who would be harmed if the dam were to fail. Since the magnitude and timing of future floods and earthquakes are indeterminate, direct determination of optimum measures to attain the economic objectives is not possible. For the same reason, simple answers to problems of equity among those affected by a dam are not available.

The probable maximum flood (PMF) has become a standard design criterion for flood protection of major dams over the past decades. The concept that equity requires that a dam impose no additional potential for damage or loss of life in downstream areas if such addition can be avoided is usually cited as the reason for this use of the PMF. Economic efficiency is not usually cited as a basis for such choice, although for large, high-hazard dams, economic considerations, if properly evaluated, possibly could lead to use of the PMF. The PMF is first of all a *hypothetical* flood based upon a set of assumptions that attempt to define the maximum flood potential for the particular site. The calculation of the PMF is based on a combination of facts, theory, and professional judgments. The methods used to calculate a PMF are not standardized, at least not to the extent that a set of individuals with the knowledge and expertise to make such calculations would independently arrive at identical evaluations. The discrepancies arise primarily from the technical, scientific, and moral issues underlying the professional judgments of the estimators as well as the lack of a quantitative definition of exactly what a PMF represents. Moral issues are involved because a dam owner may make economic decisions involving risks to others without the input or consideration of those at risk.

While it may be unsettling to accept the fact that one's ability to make estimates of the PMF is less than one might like, it would be remiss to suggest otherwise. In attempts to reassure the body politic that the level of flood for which the dam is designed is reasonable, there may have been erroneously perpetuated a myth of absolute safety by describing the PMF as one ". . . where its magnitude is such that there is virtually no chance of its being exceeded" (Bureau of Reclamation, 1981a,b), or ". . . (a flood that) . . . would have a return period approaching infinity and a probability of occurrence in any particular year, approaching zero" (Wall, 1974). Such statements suggest that the ability to predict future extreme floods is greater than that which actually exists and leads to unrealistic expectations on the part of the public. In adjudicating disputes involving claims of liability for flood damages, the courts have relied on criteria like "foreseeability" and the "appearance of certainty" to reach results that fall within the mainstream of legal analysis. However, from the perspective of the engineering profession, such concepts are of questionable merit since they do not necessarily comport with modern interpretations of probabilistic and statistical relationships.

No universal answers are available to questions on the degree of protection that should be provided for a new dam if the PMF is not the appropriate estimate of the worst possible flood that may occur, or on the actions that should be undertaken at an existing dam if new information suggests that the PMF for that structure was underestimated.

WHAT SHOULD SAFETY COST?

The responsibility for the general welfare requires that when government considers a level of protection (or safety) for its facilities, it must simultaneously consider the cost of providing that level of safety. It is faced with choosing whether an additional investment in reducing the risk to those who would be directly affected is of greater benefit to society than would be obtained by expending those funds for some other activity. In theory, it would be possible to provide extraordinarily safe dams, e.g., by providing spillway capacities equal to five or more times the PMF. Most would agree that such gross conservatism is unwarranted. Although not directly comparable, few of us would accept the idea of buying or operating automobiles that were built like army tanks even though such machines would reduce the likelihood of being injured or killed in an auto accident. For most individuals the cost and inconvenience of such protection would be viewed as excessive. We recognize differences in personal reactions to an imposed risk (e.g., the risk arising from a dam upstream from our residence) and a voluntary risk, as that imposed by our ownership of an automobile.

The way society approaches the question of risk has been dominated by feasible or practical concerns. Government by its own actions and by using its authority to regulate the behavior of individuals has generally satisfied the public desire for increased safety. It has not, however, provided explicit target levels for acceptable risk as matters of public policy. It is obvious that these questions can only be resolved through the political process.

What individuals or groups ought to demand in terms of "safety" is not entirely a technical or scientific issue. On the one hand, individuals make judgments about their participation in "voluntary" risks and may influence the riskiness of various goods and services by their willingness to pay for more (or less) safety. On the other hand, society is called upon to provide various goods and services for collections of individuals where, acting as the agent for these collective interests, government has the obligation to decide what level of risk to accept for these ("involuntary" risk) situations. In its struggles to resolve these dilemmas, government is required to consider, evaluate, and then choose among alternative courses of action that satisfy its responsibility to individuals directly affected and simultaneously the "general welfare" of society. It is unlikely that government can develop an ubiquitous risk policy with respect to all activities, because the character and consequences of the risks imposed are so different. In this sense, government cannot be expected to behave differently than individuals. Devising appropriate policies for assessment and management of risk has become a dominant dilemma of this decade.

REASONABLE CARE AND PRUDENCE IN DAM DESIGN

For the reasons discussed above, selecting the amount of protection from floods or earthquake resistance that should be included in the design of a dam is in the final analysis a matter of judgment. In order to determine whether such judgments reflect reasonable care and prudence (i.e., the exercise of good judgment and common sense) depends upon an understanding of what these criteria mean.

The floods considered here result from natural precipitation without human influence or intervention. Thus, an important premise is that certain floods are ". . . so extraordinary and devoid of human agency that reasonable care would *not* avoid the consequences . . ."—and are sometimes referred to as an act of God (New Columbia Encyclopedia, 1975). While the phrase, *act of God*, may imply to some the idea of divine intervention, it also conveys important secular concepts. First, it suggests that certain natural forces may result in events of such enormous force or consequence that no reasonable persons would plan or conduct their lives in ways that anticipate such events. Second, such acts or events occur only rarely—so infrequently that one intuitively assumes that the risk of such an event has little if anything to do with the reality of day-to-day affairs.

The understanding of nature has improved and expanded through the sciences, and engineers have been successful in applying this knowledge to the construction of facilities that provide society with some degree of protection from natural events. Thus, the hydrological, meteorological, and geological sciences have improved our understanding of extreme floods and earthquakes and provided some tools that enable us to predict limiting magnitudes for such events.

The tools for predicting floods are based upon two different scientific principles: historical observation and causality. An historical record of precipitation and runoff events (if sufficiently long) permits the use of the laws of probability and statistics to estimate the risk of floods of various magnitudes. Regardless of the accuracy of these predictive tools, they do not offer any guidance on the level of flood risk that is appropriate for any dam. That is, even if accurate predictions of the probabilities of all sizes of floods were possible, it still would be necessary to decide whether spillways of dams should be built to withstand the flood that arises once in a thousand years, in ten thousand years, in a million years, etc.

Another method of defining the flood potential at a site involves constructing a hypothetical but plausible storm (probable maximum precipitation) that is assumed to occur over a particular drainage basin where a dam exists or is contemplated. From the present knowledge of the meteorology of such storms, the geological and hydrological characteristics of the drainage basin, and their interrelations, it is possible to estimate a flood (probable maximum flood—PMF) resulting from this storm. This method of flood estimation, which is in general use, also presents a number of difficulties. First, inasmuch as the method is hypothetical, it is difficult to estimate the risk or probability of such a flood actually occurring. Second, if larger storms are observed sometime in the future (i.e., larger than the estimated probable maximum precipitation (PMP) used to calculate the PMF), these will in turn result in a bigger estimate of the probable maximum flood. (Flood estimates based on probability studies also tend to increase as more streamflow data become available.) Such an increase in probable maximum flood estimate challenges the adequacy of the existing spillway. Since this method does not provide an estimate of the risk for the original PMF, it cannot be used to determine how much additional protection would be obtained from expanding the spillways to handle a larger PMF. Obviously, the risk of dam failure due to floods will be reduced by a design that permits larger floods to pass, but such decisions must also meet the test of reasonable care and prudence.

What constitutes reasonable care and prudence in selecting the magnitude of a flood for which a dam should be designed? There appears to be no completely satisfactory answer to this question—leastwise one that would satisfy everyone. Those who would be directly affected by the possibility of a dam failure would surely choose to make the dam as floodproof as possible. Yet it is doubtful whether these individuals would be willing to pay the costs required to decrease the risk of the dam failing if the risk of failure were already relatively low (Thaler and Gould, 1982).

The current procedures used for selecting the spillway design flood (SDF) attempt to delimit reasonable care by acknowledging that the level of protection provided should reflect consideration of the hazard potential of the dam (viz., loss of human life, property damage, dam services, opportunity costs). The PMF, in spite of the fact that it is a hypothetical event of unknown risk or probability, appears to meet a standard of reasonable care, as demonstrated by the performance of dams over the past five decades. On the other hand, since the spillways of many existing dams are inadequate by PMF standards but have survived in spite of this inadequacy, it is legitimate to question whether this standard is higher than may be required. It is axiomatic that excess protection, i.e., that capacity provided at oftentimes considerable cost—but that is never used—is rarely challenged as unreasonable.

However, if a dam should fail, one can be assured that a careful inquiry will be made to determine whether the designers used reasonable care in selecting a design flood. While accountability for dam designers is essential, it is obvious that the ambiguity imbedded in the requirement to exercise reasonable care leads designers to act more conservatively in selecting a spillway design flood.

While balancing risks and costs is the ideal, this balancing cannot be accomplished with confidence at this time. In particular, the probability or average recurrence intervals of extreme floods and earthquakes can only be estimated approximately. While moving toward the ideal of balancing, some recommendations are needed to answer current concerns for the design of new dams and the retrofitting of existing dams.

The PMF is a concept that has prevented dam failures throughout the world. To a lesser extent perhaps, the similar concept of a limiting earthquake magnitude (the maximum credible earthquake—MCE) has provided a basis for preventing dam failures resulting from seismic events. To date, these notions have proven highly conservative, since few natural events have challenged them. The concepts have great usefulness in the design of new high-hazard dams. Even here, once the PMF and MCE have been estimated and the dam designed, there ought to be exploration of the cost of meeting somewhat different design levels. For example, if only a tiny addition to the cost of building the dam would be required to design to a higher standard, this greater standard makes sense. Similarly, if the cost of designing to the PMF and MCE are very large in relation to a slightly less stringent design, careful consideration must be given to whether the more stringent design is needed.

For dams that pose no threat to life, a balancing of the risks of property damage and loss of dam services against the costs of greater dam safety is appropriate. Such balancing is reasonable, since the relevant floods would be sufficiently frequent that their probabilities could be estimated with confidence.

For dams that involve a small risk to life, balancing is similarly appropriate, although the risk to even a small number of people should call forth somewhat greater safety than the case of only risk to property. Such low-hazard dams provide opportunities for research on balancing and also provide test cases for implementing the technique of balancing. The methods developed for these cases may help bring the technique of balancing into the design of high-hazard dams.

3
Summary of Present Practices on Dam Safety Standards

INVENTORY OF CURRENT PRACTICES

In preparation for this study, inquiries regarding current practices relating to safety provisions for the hazards to dams from extreme floods and earthquakes were directed to the federal agencies most concerned with dams, the appropriate unit of each state government, several private engineering firms with worldwide prominence in the dam design field, other professional organizations with interests in dam safety, and a cross section of utility firms and other organizations that own dams. Also, there have been reviews of a number of standards or policy statements, issued by technical societies relating to the safety of dams against extreme floods and earthquake hazards. Responses to these inquiries and the pertinent actions of the technical societies are summarized in Appendixes A and B.

The data from 10 federal organizations, 35 state and local agencies, 9 private firms, and 4 professional engineering societies provide a comprehensive overview of current practices in the United States and, to a great extent, in foreign countries. Because U.S. engineering firms are active in engineering for dams in other countries and because U.S. engineers play a prominent role in such organizations as the International Commission on Large Dams (ICOLD), many practices followed in the United States, particularly those of major federal dam-building agencies, have been adopted in other countries. However, there is considerable variation in the criteria adopted in the United States for evaluating the ability of dams to withstand extreme floods, especially in criteria for the smaller, less-hazardous dams.

15

CLASSIFICATION OF DAMS

Basic to all safety standards relating to hydrologic and seismic events are systems for classifying dams according to the probable damages caused by dam failure. As indicated by material in Appendixes A and B, there is considerable variety in the classification systems that have been adopted, and this variety often makes difficult any precise comparisons between criteria used by different agencies.

Most systems for classifying dams specifically utilize dam height, volume of water impounded, and character of the development in the relevant downstream area as parameters in regard to probable effects of dam failure. The classifications used by the U.S. Army Corps of Engineers in the National Dam Inspection Program are typical of such systems, and for ease of reference, the tables used in that system are shown below (Table 3-1). Although the committee has not specifically recommended a system for classifying hazard potentials, usage of the terms "low," "significant," and "high" in this report when referring to hazards generally conforms to Table 3-1, with the term "intermediate" being used interchangeably with "significant."

A number of federal and state agencies (e.g., U.S. Forest Service, Alaska, Illinois, South Carolina, and Virginia) have adopted a classification system

TABLE 3-1 Terms for Classifying Hazard Potentials

Category	Impoundment (ac-ft)	Height of Dam (ft)
Size of dam[a]		
Small	50 to 1,000	25 to 40
Intermediate	1,000 to 50,000	40 to 100
Large	Over 50,000	Over 100

Category	Loss of Life (Extent of Development)	Economic Loss
Hazard potential classification		
Low	None expected (no permanent structures for human habitation)	Minimal (undeveloped to occasional structures or agriculture)
Significant	Few (no urban developments and no more than a small number of inhabitable structures)	Appreciable (notable agriculture, industry, or structures)
High	More than few	Excessive (extensive community, industry, or agriculture)

[a]Criterion that places project in largest category governs.

either identical to or essentially the same as that displayed below. Other agencies and states (e.g., U.S. Soil Conservation Service, Arizona, Kansas, Missouri, and Pennsylvania) have systems that are similar but use the product of dam height (in feet) and storage volume (in acre-feet) as a size criterion. A number of states (e.g., Arizona, Arkansas, Georgia, Kansas, North Carolina, North Dakota, and South Carolina) have four or more categories of size. Arkansas appears to base its dam classification entirely on storage capacities and drainage areas, while Georgia and North Dakota utilize only height of dam and reservoir storage in their systems. New Jersey has three categories based on "hazard potential" and one labeled "small dams." The classification systems of the Tennessee Valley Authority, the State of Utah, and the Institution of Civil Engineers of the United Kingdom in general refer only to the level of hazards that would be created by failure of the dam.

In some systems for classifying dams, overall evaluation of the factors affecting downstream hazards is implied, but criteria for such evaluation are not set out. The following types of classification criteria, described in Appendix A, are such systems:

- "loading conditions" as used by the U.S. Bureau of Reclamation,
- "functional design standards" as used by the U.S. Army Corps of Engineers, and
- "security standards" as used by the U.S. Committee on Large Dams (1970).

While it appears that many of the differences in dam classification systems are the result of arbitrary choices of regulatory authorities, it also appears that most of the classification systems have been structured to meet the perceived needs of the issuing agency or state government.

SPILLWAY CAPACITY CRITERIA

Table 3-2 shows all the spillway capacity criteria as stated in agency standards in terms of either design rainfall or design floods reported to be in current use by the entities responding to the committee's inquiries. Criteria based on estimates of probable maximum precipitation (PMP), and probable maximum flood (PMF) are widely used. In fact there is some indication that corresponding values (e.g., 0.50 PMP and 0.50 PMF) are used more or less interchangeably by some engineers. The mixed criteria are listed by the Soil Conservation Service and by West Virginia. The California Division of Safety of Dams allows use of the one-in-1,000-year flood as the required minimum flood for spillway design for low hazard dams. Michigan's criteria call for use of a 200-year flood. Those two are the only references to any

TABLE 3-2   Spillway Capacity Criteria Reported to Be in Current Use by Various Agencies

Deterministic Criteria	Mixed Criteria	Probabilistic Criteria
Criteria specifying rainfalls		
PMP	$P_{100} + 0.40 (PMP-P_{100})$	$2.25\,P_{100}$
0.90 PMP	$P_{100} + 0.26 (PMP-P_{100})$	$1.50\,P_{100}$
0.80 PMP	$P_{100} + 0.12 (PMP-P_{100})$	P_{100}
0.75 PMP	$P_{100} + 0.06 (PMP-P_{100})$	P_{50}
0.50 PMP		P_{10}
0.45 PMP		
0.40 PMP		
0.33 PMP		
0.30 PMP		
0.25 PMP		
0.225 PMP		
0.20 PMP		
0.10 PMP		
Criteria specifying floods		
PMF		10,000
0.75 PMF		1,000
0.50 PMF		200
0.40 PMF		150
0.30 PMF		100
0.25 PMF		50
0.20 PMF		

NOTES: This is simply a listing of reported criteria. Position of entries in adjacent columns does not imply any relationship. PMP, probable maximum precipitation; PMF, probable maximum flood; P (with subscripts), precipitation having average return period in years indicated by subscript; 10,000 year, etc., flood having indicated average return period.

average return period greater than 100 years in any criteria in use in the United States. The 10,000-, 1,000-, and 150-year frequency floods are listed in the criteria of the Institution of Civil Engineers, London.

Table 3-3 gives an approximate comparison (based on the classifications used for the National Dam Inspection Program) of the various criteria more fully described in Appendix A. As noted above, differences in systems for classifying dams make precise general comparison of this type difficult. Although not set out in many published criteria, there appears to be growing use of dam safety evaluation procedures based on estimating effects in relevant downstream areas of a dam being overtopped and failing during vari-

TABLE 3-3 Comparison of Indicated Spillway Capacity Criteria in Use or Proposed

Hazard Class:	High			Significant			Low		
Size of Dam:	Large	Inter-mediate	Small	Large	Inter-mediate	Small	Large	Inter-mediate	Small
Federal agencies									
Ad Hoc ICODS of FCCSET	PMF	PMF	PMF	PMF	PMF	PMF	*	*	*
Bureau of Reclamation	PMF	PMF	PMF	*	*	*	*	*	*
FERC	PMF	PMF	PMF	PMF	PMF	PMF	*	*	*
Forest Service	(See Corps criteria for National Dam Inspection Program								)
ICODS	PMF	PMF	PMF	*	*	*	*	*	*
National Weather Service				Does not establish criteria for dams					
SCS	PMP	PMP	PMP	($P_{100}+0.4(PMP-P_{100})$		)	*	*	*
TVA	PMF	PMF	PMF	(TVA Max. Prob. Fld.		)	*	*	*
Corps of Engineers (Corps Projects)	PMF	PMF	PMF	*	*	*	*	*	*
Corps of Engineers (National Inspection Program)	PMF	PMF	1/2 PMF to PMF	PMF	1/2 PMF to PMF	100 yr to 1/2 PMF	1/2 PMF to PMF	100 yr to 1/2 PMF	50 yr to 100 yr
Nuclear Regulatory Commission	PMF	PMF	PMF	(See Corps criteria for National Dam Inspection Program					)
State agencies									
Alaska	(See Corps criteria for National Dam Inspection Program								)
Arizona	PMF	1/2 PMF to PMF	1/2 PMF	1/2 PMF to PMF	1/2 PMF	100 yr to 1/2 PMF	1/2 PMF to PMF	100 yr to 1/2 PMF	100 yr

TABLE 3-3 (Continued)

Hazard Class:	High			Significant			Low		
Size of Dam:	Large	Inter-mediate	Small	Large	Inter-mediate	Small	Large	Inter-mediate	Small
Arkansas	*	*	*	*	*	*	*	*	*
California	PMF	PMF	*	*	*	*	*	*	1,000 yr
Colorado	PMF	*	*	*	*	*	*	*	100 yr
Georgia	PMP	50% of PMP	33% of PMP	*	*	*	*	*	*
Hawaii	Does not have authorized dam safety program								
Illinois	PMF	PMF	1/2 PMF to PMF	PMF	1/2 PMF to PMF	100 yr to 1/2 PMF	1/2 PMF to PMF	100 yr to 1/2 PMF	100 yr
Indiana	No criteria for dams furnished								
Kansas	0.4 PMP	0.3 PMP	0.25 PMP	0.4 PMP	0.3 PMP	100 yr	0.4 PMP	0.25 PMP	100 yr
Louisiana			PMP	Proposed, see Corps criteria for National Dam Inspection Program					
Maine	*			Regulations to be developed					
Michigan	*	*	*	*		*	*	*	*
Mississippi (new dams)					See SCS criteria				
Mississippi (existing dams)	0.5 PMF	0.5 PMF	0.5 PMF	*		*	*	*	*
Missouri	*	*	*	*		*	*	*	*
Nebraska	"Relatively same as used by Federal agencies"								
New Jersey	PMP	PMP	PMP	1/2 PMP	1/2 PMP	1/2 PMP	100 yr	100 yr	100 yr
New Mexico				See SCS criteria					
New York (new dams)	PMF	*	0.5 PMF	0.4 PMF		225% of 100 yr RF	150% of 100 yr RF		100 yr
New York (existing dams)	1/2 PMF	1/2 PMF	1/2 PMF	150% of 100 yr	150% of 100 yr		100 yr	100 yr	100 yr
North Carolina	PMP	3/4 PMP	1/2 PMP	3/4 PMP	1/2 PMP	1/3 PMP	100 yr	1/3 PMP	100 yr
North Dakota	PMP		1/2 PMP	3/4 PMP	1/2 PMP	1/3 PMP	1/2 PMP		100 yr
	Being developed—anticipate use of Corps criteria for National Dam Inspection Program								

Ohio	PMF	PMF	0.5 PMF	0.5 PMF	0.5 PMF	0.25 PMF	0.25 PMF	0.25 PMF
Pennsylvania	See Corps criteria for National Dam Inspection Program							
South Carolina	See Corps criteria for National Dam Inspection Program							
South Dakota	No state dam safety program							
Texas	Corps and SCS criteria have been used—new criteria being developed							
Utah	½ PMF to PMF		100 yr to ½ PMF			100 yr	100 yr	100 yr
Virginia	See Corps criteria for National Dam Inspection Program							
Washington	Criteria based on rainfall frequencies under development							
West Virginia	PMP	PMP	$P_{100} + 0.4(PMP-P_{100})$			$P_{100} + 0.12(PMP-P_{100})$		
Other government agencies								
City of Los Angeles	See criteria of State of California							
East Bay Municipal Utility District, California	Uses "state-of-the-art criteria"							
Salt River Project, Arizona	See Bureau of Reclamation criteria							
South Carolina Public Service Authority	See FERC criteria							
Technical societies								
American Society of Civil Engineers (ASCE)	PMF	PMF	PMF	PMF	PMF	PMF	*	*
International Commission on Large Dams (ICOLD) (draft guidelines)	PMF	PMF	PMF	PMF	PMF	*	*	*
U.S. Committee on Large Dams (USCOLD)	Has not promulgated criteria							

TABLE 3-3 (Continued)

Hazard Class:	High			Significant			Low		
Size of Dam:	Large	Inter-mediate	Small	Large	Inter-mediate	Small	Large	Inter-mediate	Small
U.S. firms									
Acres American, Inc.									
Alabama Power Co.	Designs spillways for 10,000-yr flood, checks design with maximum probable flood — See FERC criteria								
R. W. Beck and Assoc.	Generally follows Corps of Engineers criteria								
Central Main Power Co.	See FERC criteria								
Duke Power Co.	See FERC criteria								
Charles T. Main, Inc.	Uses PMF for major dams								
Planning Research Corporation	Uses PMF for significant and high-hazard projects. Has based spillway design on dam break analyses.								
Yankee Atomic Electric Co.	See criteria of Nuclear Regulatory Commission								
Other U.S. entities									
Illinois Association of Lake Communities	Furnish no standards, but protested any requirement that existing dams meet new dam safety criteria. Questioned legality of such requirement.								
Foreign									
Institution of Civil Engineers, London	PMF	PMF	PMF	(0.5 PMF or 10,000-yr fld, take larger)			(0.3 PMF or 1,000 yr, take larger)		

*Not specifically defined, or agency criteria not comparable to classifications used in this table.

NOTE: Criterion shown in parentheses applies to all classifications within the parentheses.

ous size floods. By such procedures, the dam safety evaluation flood selection is based on these estimated effects. Thus, fixed criteria, such as illustrated in Tables 3-2 and 3-3, are not used.

This inventory of current practices in providing dam safety during extreme floods shows considerable diversity in approach by various federal, state, and local government agencies, professional societies, and privately owned firms. There is a fair consensus on the spillway requirements for large, high-hazard dams. But the results of the inventory show widespread uncertainty as to what might be appropriate hydrologic criteria for safety of other classes of dams. From study of the inventory results, the following observations can be made:

- Use of PMP estimates for evaluating spillway capacity requirements for large, high-hazard dams predominates, although a number of state agencies have indicated that their standards do not require that such dams pass the full estimated PMF based on the PMP.
- The influence of the practices of the principal federal dam-building agencies is evident in the majority of the standards for large, high-hazard dams, but the practices of those agencies have had less effect on current state standards for small dams in less hazardous situations.
- Apparently as a result of the National Dam Inspection Program for nonfederal dams carried out by the Corps of Engineers in the 1977-1981 period, several state dam safety agencies have adopted the spillway capacity criteria used in those inspections.
- Several states have adapted the standards used by the Soil Conservation Service for the design of the tens of thousands of smaller dams constructed under that agency's programs.
- Current practices include use of arbitrary criteria (such as 150 percent of the 100-year flood, fractions of the PMF, and combinations of the PMF with probability based floods) for which there is no apparent scientific rationale.
- Practices of the major federal dam-building agencies for large, high-hazard dams have been adopted by most U.S. companies owning dams and by U.S. engineering firms designing dams for domestic and foreign clients. (The regulations of the Federal Energy Regulatory Commission have required such standards for licensed hydroelectric projects.)
- It appears that only three agencies (the Federal Energy Regulatory Commission, the Mississippi Department of Natural Resources, and the New York State Department of Environmental Conservation) have issued explicit standards for existing dams that differ from the requirements for new dams. (However, other responses did not specifically state whether different standards were applicable to existing dams.)

CRITERIA FOR EARTHQUAKE EFFECTS

Table 3-4 shows a summary of current practices in evaluating safety of dams against earthquakes as specifically reported in response to the committee's requests for information. Since many of the responses gave no specific information in regard to a number of the practices shown, this table should be used with caution, as some of the agencies may be actually using more of the approaches to dam safety than indicated. However, even though it may be not completely reliable, Table 3-4 does give some indication of the probable extent of use of the various techniques for analysis of earthquake effects on dams.

Seismic Zones of the United States

Seismic zone maps of the United States (Algermissen, 1969; Algermissen and Perkins, 1976) are used by most federal and state agencies as basic references when deciding if any seismic factors should be considered in dam design and if special investigations are required. Since such maps are incorporated in most building codes, they are often employed in selecting design criteria for buildings and ancillary structures and systems at dams. The zone maps, in part, are developed on the basis of historic seismicity, such as shown in Figure 3-1. The historic earthquake record indicates that damaging earthquakes occur throughout the United States. Figure 3-2 shows the Algermissen (1969) seismic zone map, as it appears in the *Uniform Building Code, 1979 Edition*.

The Soil Conservation Service, the Corps of Engineers, and some states, when dynamic response analyses are not required, employ seismic zones for determining the minimum seismic coefficients for pseudostatic analyses. The following Corps of Engineers criteria (giving coefficients to be multiplied by weight of structure to determine estimated horizontal earthquake loadings) are typical:

Seismic Zone	Minimum Coefficient ($\times$ g)
0	0
1	0.05
2	0.10
3	0.15
4	0.20

TABLE 3-4 Summary of Practices Specifically Reported Relating to Evaluation of Safety of Dams Against Earthquake Hazards

Agency (1)	Practices or Techniques Used or Required						
	Seismic Risk Zone Maps	MCE	OBE	Analyses of Liquefaction Potentials	Pseudo-Static Analyses	Dynamic Response Analyses	Defensive Design Measures
Federal agencies							
Ad Hoc Interagency Comm. of FCCSET		*					
Bureau of Reclamation (2) (3)		*	*				
FERC (2)		*	*				
Forest Service (4)							
ICODS (2)		*	*				
SCS	*	*	*	*	*		
TVA		*	*	*	*	*	
Corps of Engrs (5)		*	*	*	*	*	
Nuclear Regulatory Commission (2)		*	*				
State Agencies							
Alaska	*						
Arizona (1)	*		*	*			
California		*			*		
Colorado		*		*	*	*	*
Illinois	*				*	*	
Kansas	*						
Mississippi (6)							*

TABLE 3-4 (Continued)

Agency (1)	Practices or Techniques Used or Required						
	Seismic Risk Zone Maps	MCE	OBE	Analyses of Liquefaction Potentials	Pseudo-Static Analyses	Dynamic Response Analyses	Defensive Design Measures
Missouri	*						
Nebraska (6)							
New Jersey (6)	*				*		
New Mexico	*				*		
New York	*				*		
North Carolina					*		
Utah (7)	*	*	*				
West Virginia				*	*		
Other government agencies							
City of Los Angeles		*		*		*	*
East Bay Municipal Utility District		*		*		*	
New York Power Authority				*		*	
Salt River Project, AZ		*					
South Carolina Public Service Authority (8) (Santee-Cooper)							
U.S. firms							
Acres American, Inc.		*		*	*	*	
Alabama Power Co. (8)					*		*

Central Maine Power
  Co. (8)
Duke Power Co. (8)
Charles T. Main,
  Inc. (6)
Planning Research
  Corp. (2)
R. W. Beck and
  Assoc.

*Practice reported to be in use or required.

NOTES:

(1) The following either reported that no formal standards for evaluation of earthquake hazards have been adopted or failed to furnish such standards (in some cases data on informal practices were furnished): Arizona, North Dakota, Arkansas, Ohio, Georgia, Pennsylvania, Hawaii, South Carolina, Indiana, South Dakota, Louisiana, Texas, Maine, Virginia, Michigan, Washington.

(2) Uses different term for MCE or OBE.

(3) Bureau of Reclamation considers reservoir-induced seismicity.

(4) Forest Service requires evaluation of earth movement potentials and establishment of design criteria on case-by-case basis.

(5) Practices shown apply to U.S. Army Corps projects. In the phase I investigations for the National Dam Inspection Program, preliminary assessments of potential vulnerability to seismic events were required.

(6) Follows practices of Soil Conservation Service or U.S. Army Corps of Engineers or Bureau of Reclamation.

(7) Requires consideration of reservoir-induced seismicity, seismic waves, and instability of reservoir slopes and bottom.

(8) Follow FERC guidelines.

28

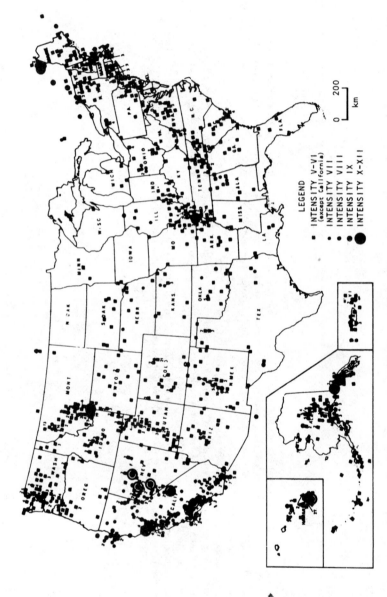

FIGURE 3-1   Earthquakes with maximum Modified Mercalli intensities of V or above in the United States and Puerto Rico through 1976. Source: Algermissen (1983).

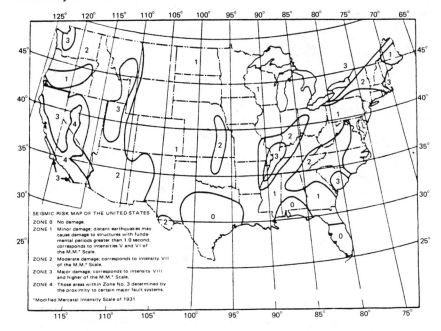

FIGURE 3-2 Seismic risk map of the United States. Source: Reproduced from the *Uniform Building Code, 1979* (1982) (1985) edition, © 1979 (1982) (1985), with permission of the publisher, the International Conference of Building Officials.

Earthquake Intensity and Magnitude Scales

Both the Modified Mercalli intensity scale (Table 3-5) and the Richter magnitude scale are in use to describe earthquakes, although dam safety criteria usually refer to the Richter scale.

The Richter magnitude scale describes the size of the earthquake; that is, it describes the seismic energy released by a fault rupture as well as the size of the area affected by strong ground shaking. Thus, Richter magnitudes are not directly comparable with Mercalli intensity ratings that vary over the affected area. The approximate relationships between Richter magnitudes and areas in square miles affected by various levels of peak accelerations, as developed for the western United States, are shown in Table 3-6.

Seismologists actually have defined four types of magnitude measures or scales: M_L, M_S, m_b, and M_w, which are based on the recorded amplitudes of local waves, surface waves, body waves, and very long period waves, respectively; these waves become prominent on seismograms at different distances from the earthquake source. The Richter magnitudes commonly reported in the news media are actually M_L for earthquakes relatively close to the seismograph and M_S for earthquakes at greater distances from the seismograph.

TABLE 3-5   Modified Mercalli Intensity Ratings

MMI	Condensed Description
I	Not felt
II	Felt indoors by few, especially on upper floors
III	Felt indoors by several
IV	Felt indoors by many, outdoors by few
V	Felt indoors by practically all, outdoors by many
VI	Felt by all, damage slight in poorly built buildings
VII	Damage negligible in buildings of good design and construction, slight to moderate in well-built ordinary buildings, considerable in poorly built or badly designed buildings
VIII	Damage slight in structures built especially to withstand earthquakes, considerable in ordinary substantial buildings
IX	Damage considerable in structures built especially to withstand earthquakes
X	Many specially designed structures destroyed
XI	Few, if any, structures remaining standing
XII	Complete destruction

The commonly used Richter magnitude (based on M_L for smaller earthquakes and M_S for large earthquakes) does not have the same numerical value as the m_b and M_w; hence these differences should be taken into account.

Seismic Design Terminology

The following terms in reference to ground motions at the dam site, or to earthquakes causing those motions, are presently used by various government agencies and other entities with respect to seismic design criteria:

DBE—design basis earthquake (Planning Research Corporation)
DE—design earthquake (R. W. Beck & Associates)
EDBE—economic design basis earthquake (USBR, California)
MCE—maximum credible earthquake (most agencies)
MCGM—maximum credible ground motion (USBR, California)
MDE—maximum design earthquake (ICODS)
ME—maximum earthquake (FERC)
OBE—operational basis earthquake (NRC, USACE, TVA)
PMA—probable maximum acceleration (Missouri)
SSE—safe shutdown earthquake (NRC)

Definitions of these terms as used by various entities follow. Maximum credible earthquake (MCE) is defined by the U.S. Army Corps of Engineers

TABLE 3-6 Richter Magnitudes and Square-Mile Areas as Affected by Peak Accelerations

Peak Acceleration (percent of acceleration of gravity equalled or exceeded)	Earthquake Magnitude (Richter) (area in square miles)			
	5	6	7	8
5	400	3,600	13,000	50,000
10	90	1,600	7,500	30,000
20	–	150	2,500	14,000
30		–	300	6,000
40			–	3,000
50			–	–

as, "the earthquake that would cause the most severe vibratory ground motion or foundation dislocation capable of being produced *at the site* under the *currently known* tectonic framework" (emphasis supplied); by the U.S. Bureau of Reclamation as, "at a specific seismic source the maximum earthquake that appears capable of occurring in the presently known tectonic framework. It is a rational and believable event that is in accord with all known geologic and seismologic facts"; by the Tennessee Valley Authority as, "the earthquake associated with specific seismotectonic structures, source areas, or provinces that would cause the most severe vibratory ground motion or foundation dislocation capable of being produced at the site under the currently known tectonic framework"; and by the Interagency Committee on Dam Safety as, "the hypothetical earthquake from a given source that could produce the severest vibratory ground motion at the dam."

The term maximum credible ground motion (MCGM) or equivalent terminology is used by the U.S. Bureau of Reclamation, California Division of Safety of Dams, and others; however, all entities do not define MCGM in the same way for use in dam safety analysis. For such purpose, MCGM may be described by several of the following sets of data:

- peak ground acceleration
- peak ground velocity
- duration of strong shaking
- response spectrum
- time history (a recording of ground acceleration versus time)

The draft paper entitled "Proposed Federal Guidelines for Earthquake Analysis and Design of Dams," developed by an interagency task group of the Interagency Committee on Dam Safety (ICODS), uses the term maximum design earthquake to describe the earthquake selected for design analysis of a project after considering earthquake potential of the dam site and the potential losses from failure of the dam.

The term maximum earthquake used by the Federal Energy Regulatory Commission, apparently generally refers to the same type event as the term maximum credible earthquake.

Planning Research Corporation has defined design basis earthquake (DBE) as, "the largest earthquake which would be expected to occur once during the expected life of the project."

R. W. Beck & Associates used design earthquake (DE) in the design of Swan Lake arch dam in Alaska, and define it as the "largest earthquake that would be expected to occur during the economic life of the dam (recurrence interval of once in 100 years)." "Largest earthquake" implies the "earthquake producing the greatest loading on the structure."

The U.S. Bureau of Reclamation defines the economic design basis earthquake (EDBE) as that earthquake under the loading from which "the project facilities not critical to the retention or release of the reservoir would be designed to sustain the earthquake with repairable damage." The degree of damage that would be acceptable could be based on an economic analysis or an estimate of the cost of the repair versus the initial cost to repair the damage.

Operating basis earthquake is defined by the U.S. Army Corps of Engineers as, "the maximum level of ground motion that can be expected to occur at the site during the economic life of the project, usually 100 years"; by the Tennessee Valley Authority as, "the earthquake for which the dam is designed to resist and remain operational"; and by the Nuclear Regulatory Commission as, "that earthquake which, considering the regional and local geology and seismology and specific characteristics of local subsurface material, could reasonably be expected to affect the plant site during the operating life of the plant."

Regulations proposed by the State of Missouri specify a fraction of probable maximum acceleration (PMA) as the design acceleration for various stages of design and different classes of dams, defining PMA as the "probable maximum acceleration of bedrock determined by the seismic zones" used by the U.S. Army Corps of Engineers.

The Nuclear Regulatory Commission defines safe shutdown earthquake (SSE) as that earthquake based on an evaluation of the maximum earthquake potential considering the regional and local geology and specific characteristics of local surface material.

Maximum credible and safe shutdown earthquakes are used to evaluate the safety of dams; however, some damage to the facility during such an earthquake is acceptable, provided there is no release of reservoir water. All respondents said they determine such earthquakes for dams using deterministic approaches, except Acres American, Inc., which develops both probabilistic and deterministic approaches and adopts the most severe earthquake found by either approach. Operating basis, design, design basis, and economic design basis earthquakes are used for dynamic analyses. Probabilistic methods sometimes are used to develop these earthquake estimates. Minimal damage is expected from these earthquakes.

Seismic Criteria for Pseudostatic Stability Analysis

Before the mid-1960s only the pseudostatic methods of stability analysis were used for dams in seismic areas. The seismic load was assumed to be a sustained horizontal force acting on the dam in the most critical direction. Depending upon the size of the dam and the seismic risk, the seismic force was assumed to range from 0.05 to 0.15 times the weight of the structure.

For larger dams the U.S. Bureau of Reclamation combined horizontal acceleration effects with a vertical component, which was 50 percent of the horizontal acceleration; the assumed directions of the two components were those most unfavorable to structural stability. Most large foreign dams adopted similar criteria. For example, Bhakra Dam in India, a 740-foot-high concrete gravity structure located about 180 kilometers from the epicenter of the Richter magnitude (M) 8.6 Kangra earthquake of 1905, was designed for a lateral force coefficient of 0.15 and a vertical force coefficient of 0.075. The U.S. Army Corps of Engineers still requires the use of seismic coefficients for sliding and stability analyses of concrete dams and structures. Hydrodynamic pressures also were taken into account by similar methods in some cases.

Dynamic Response Analyses

Analysis of the dynamic response of a dam to specified earthquake ground motion, when it is located in seismic zones 3 or 4 (and under some conditions, in zone 2), is now part of the dam safety criteria of most federal agencies. Only three states specifically stated that dynamic response analyses were used; others stated they used standards of federal agencies and, thus, implied use of such analyses. It is surmised that states that have relatively new dam safety programs and/or do not have mapped active faults within their boundaries are using pseudostatic methods or are not considering earthquake loadings.

The dynamic response analyses may be two- or three-dimensional and employ the finite element technique. Depending on the size and type of dam, the foundation characteristics, and the severity of the design earthquake, the analysis may consider deformations of the foundations and abutments as well as the structure.

For embankment dams, the principal objectives of a dynamic analysis are assessment of liquefaction potential of susceptible materials and determination of permanent deformations and the potential for cracking. For concrete dams, the dynamic response analyses determine the instantaneous total (dynamic plus static) stresses at both faces of the dam and at designated locations.

The possible effects of fault movement on the dam are included as appropriate.

4
History of Development of Present Practices

DESIGN CRITERIA FOR EXTREME FLOODS

Engineers designing dams have been and continue to be handicapped by lack of reliable bases for estimates of future extreme flood runoffs from the basins upstream from their dams.

Myers (1967) has described the evolution in the United States of the current practices for estimating extreme flood-producing capabilities of watersheds for use in spillway design. The main developments in that evolution took place in a 30-year period, from about 1910 until 1940. The central concepts in the current methods of estimating spillway requirements have changed little in the last few decades, but there have been many refinements in their application. Also, in the last 40 years, the concepts and methodology developed in the United States have become fairly well established throughout the world as the standard approach for sizing spillways of large dams where failure of the dams would result in hazards to life and property.

As noted by Myers, the evolution in the United States of techniques for determining spillway capacity requirements can be divided into four stages, outlined below.

Early Period

Before 1900 the designer of a dam in the United States usually had little hydrologic data on which to base estimates of spillway requirements. The rainfall reporting system of the U.S. Weather Bureau was largely a widely

35

scattered system of nonrecording gages that reported daily rainfall amounts primarily in the interest of agriculture. Systematic collection of stream stage data was being accomplished at a relatively small number of gages on major streams. Often the designer of a dam had little information except high-water marks on the stream he was damming or in adjacent watersheds to indicate the flood potentials at his dam site. The designer might estimate peak discharge rates of past floods based on such meager data and base his spillway design on such estimates with, perhaps, some added capacity provided as a safety factor. Some spillways were designed for some multiple of the maximum known flood, e.g., twice the maximum known flood. Myers noted a tendency among engineers of this era to assume that nature had already demonstrated the maximum flood potential on each stream and that spillway designs could be based with safety on the records available of past floods. Some earth dams built during this period failed because of overtopping, but since most dams of the era were relatively low masonry or rock-filled timber-crib structures, overtopping could be experienced by many structures without failure.

Regional Discharge Period

As more stream discharge data became available, engineers began to recognize that looking at all past flood peaks in a region might give more reliable estimates of maximum flood-producing potentials than a limited record on a single stream. This idea recognized the random-chance nature of major storm rainfalls occurring over a specific watershed. It also introduced the concept that hydrologic data observed at one location, when appropriately transposed, could serve as a basis for estimates at other locations. A flood-frequency formula developed by W. E. Fuller (1914), based on analysis of flood peaks in hundreds of streams, sought to relate the peak discharges having various average return periods to the mean of the highest annual floods from the same watershed. Later formulas, notably one known as "Myers" rating and others developed by W. P. Creager and C. S. Jarvis, based on enveloping the maximum observed floods from many watersheds, sought to relate the "maximum flood" from a watershed to some function of the size of the area drained. Prior to about 1940, many spillway designs for large dams were based on such formulas developed from regional analyses of maximum flood peaks. Such formulas are still used to compare observed and computed flood peaks.

Myers noted that, about the year 1930, formal statistical methods began to be applied to hydrologic problems, including estimating maximum floods. Some engineers felt that, given a reasonably long period of discharge record and the right type of frequency analysis, reliable estimates of rare flood peaks could be obtained for determining spillway design requirements. However, the occurrence of floods far exceeding any that could have been estimated by frequency analysis of past records of basin discharges has discouraged the use of frequency analysis as a tool for estimating the magnitude of rare floods. With the longest flood records in the United States extending back about 150 years, most American hydrologists are reluctant to give estimates of stages or discharges for floods with long average return periods. However, in recent years the National Flood Insurance Program has required numerous estimates of floods having annual probabilities of exceedance of 0.01, or average return period of 100 years for the purpose of defining floodplains in connection with community planning activities and administration of the flood insurance program. Also, some state dam safety agencies call for similar floods for design of spillways for small, low-hazard dams. In contrast to American practice, the guide issued by the Institution of Civil Engineers in London calls for estimating 10,000-year floods in spillway design.

Storm Transposition Period

In the 1920s engineers in the Miami Conservancy District in Ohio undertook a program of studying past major flood-producing storms to develop rainfall duration-area-depth relationships for use in planning and design of the comprehensive flood control project for the Miami Valley. It was recognized that measured flood peaks are dependent on topography and size of individual watersheds and chance placement of storm centers over the watershed. Also, within meteorologically similar areas, observed maximum rainfall values could provide a better general indication of maximum flood potentials than flood discharges from individual watersheds. In the 1930s, the work that had begun in the Miami Conservancy District was used by other engineers, notably those in the Tennessee Valley Authority and the U.S. Army Corps of Engineers, to develop a system for estimating what was termed "maximum possible" rainfalls. The independent development of the unit hydrograph concept in the early 1930s gave a way of converting the maximum possible rainfall values into streamflow to produce maximum possible spillway design floods.

The method of transposing and enveloping past maximum storm rainfalls to obtain maximum possible estimates was illustrated graphically in some U.S. Army Corps of Engineers reports of the period by what was termed a "peg" model. Suppose it was desired to determine the maximum 2-day rainfall over a 500-square-mile drainage area above a proposed dam site. From duration-area-depth data available for past major storms in the meteorologically similar area, the maximum rainfall depth over 500 square miles and 2-day duration for each relevant storm was determined. Then, on a map of the region, a vertical wooden peg whose height represented, to some convenient scale, the 500-square-mile 2-day rainfall value for each storm was placed at the geographic location where the rainfall was observed. Strings were stretched connecting the tops of pegs representing individual storms at their respective locations. The height of the network of strings over the map location of the basin at the proposed dam site represented the estimated maximum possible rainfall over the basin.

Probable Maximum Precipitation (PMP) Period

By the end of the 1930s it was recognized that the enveloping and transposing of past major storms might not necessarily yield the upper limit of probable rainfall over a basin. Concepts of air mass analysis then being introduced showed that in any storm system such factors as the humidity of the incoming air, the velocity of wind bringing moisture-laden air into the basin, and the percent of the water vapor that can be precipitated serve to place limits on the amount of precipitation from the storm. Thus, if in any major storm of record, any of the limiting factors was less than what could be expected in the basin being studied, an adjustment (i.e., increase) in the observed rainfall values would be indicated if maximum probable rainfall values were sought. Such transposition, adjustment, and enveloping techniques are the bases for the PMP estimates now in use and which are more fully discussed in Chapter 5 and Appendix C.

Composite Criteria

From the summary of present practices in Chapter 3, it will be recognized that both PMP estimates and estimates based on frequency analysis of either rainfalls or streamflows are in use for determining sizes of spillways. Some agencies specify various percentages of the PMP or the probable maximum flood (PMF) derived from the PMP as bases for spillway design. Other criteria attempt to combine flow frequency concepts and PMF concepts. Considering the great differences in the basic concepts involved, the rationale for such combinations seems questionable.

Risk-Based Analyses

Other bases for determining spillway capacity requirements have been suggested. A 1973 report of an American Society of Civil Engineers (ASCE) committee advocated that spillway capacities of existing dams be reevaluated by economic analyses of costs of spillways of various capacities and the estimated long-term damages (or risk costs) associated with each spillway capacity (ASCE, 1973). The committee proposed that costs associated with loss of life be evaluated on the basis of current practices of courts in awarding damages in cases involving accidental deaths. The spillway capacity selected would be that which gave lowest total costs (i.e., the lowest sum of risk costs plus structure costs). The ASCE committee's proposals have not been widely adopted. It appears that the following factors have contributed to the lack of enthusiasm for such an approach:

- There is a great reluctance to attempt to place a value on a human life in such a decision process.
- The results of such economic analysis of present-day costs versus possible future damages are very much dependent upon the interest rate selected for the analysis, and many engineers are reluctant to have design involving safety determined by the interest rate that happened to be applicable when the design was done. (This factor has less importance for governmental agencies in which the discount rate to be used for public sector benefit-cost studies is prescribed (based on some chosen social discount rate) and which is permitted to change very slowly from year to year).
- Estimates of average annual risk costs are dependent on the flood frequency curve adopted, and the persistent problem of estimating frequencies of rare flood events becomes a deterrent.

As discussed in Chapter 5, another risk-based approach to determining spillway capacity that has been used involves estimating, by dam break analysis and flood routings, the flood flow for which failure of the dam would cease to create significant additional damage downstream over damage sustained without failure. That flood then becomes the design discharge for the spillway. This approach has merit because it avoids unneeded expense for added spillway capacity and does not worsen the situation downstream in case of failure.

Other risk-based analysis methods similar to the method advocated by the 1973 ASCE committee report, but which do not attempt to place a value on human life, are being advocated in those agencies attempting to apply risk analysis techniques to decisions regarding dam safety. In such analyses, potentials for deaths resulting from dam failure become an element in the decision process but not part of the economic evaluation.

EARTHQUAKE-RESISTANT DESIGN OF DAMS

The possible effects of earthquakes on the safety of dams were first taken into account by the engineering profession as early as the middle 1920s. The classic paper by H. M. Westergaard entitled "Water Pressure on Dams During Earthquakes" was published in 1933. There is every indication that many design agencies were at that time making some analytical studies to evaluate seismic safety, or incorporating simple defensive measures into design of projects to increase the safety of dams against earthquake-shaking effects.

In the period 1930-1970, design practice usually considered earthquake effects by simply incorporating in the stability or stress analysis for a dam a static lateral force intended to represent the inertia force induced by the earthquake. Most often this force was expressed as the product of a lateral force coefficient and the force of gravity. The coefficient generally varied, depending on the seismicity of the area in which the dam was located and the judgment of the engineer involved, between values of about 0.05 and 0.15. This method of approach was termed the pseudostatic analysis method, in recognition of the fact that the static lateral forces were only intended to represent the effects of the actual dynamic earthquake forces, which effects could well be substantially different from those of the static forces used in the analysis procedure.

This approach was essentially similar to that used for building design in seismically active areas. For concrete dams it was customary to consider also the hydrodynamic pressures on the face of the dam using the approximate method proposed by Westergaard. For earth dams, studies by Zangar (1952) led to the conclusion that consideration of hydrodynamic pressures was generally unnecessary but might be required for dams with relatively steep slopes.

Pseudostatic analyses, combined with engineering judgment, were the only methods used to assess the seismic stability of dams until the late 1960s, and these approaches were generally considered entirely adequate by dam engineers and regulatory agencies based on the fact that the field performance of dams designed by these procedures and subsequently subjected to strong earthquake shaking was generally found to be satisfactory. The observed performances lending support to this belief included:

• The excellent performance of dams located close to the San Andreas fault in the 1906 San Francisco earthquake (M 8.3). At the time of that event there were 14 earth dams located within 5 miles of the fault on which the earthquake occurred, and none of these suffered any significant damage. There was also one concrete dam, the Crystal Springs Dam, located only a few hundred yards from the San Andreas fault that survived the earthquake with no apparent damage.

• Several 200-foot-high concrete gravity dams in Japan were also subjected to earthquake-shaking intensities of M 8 with no damage.

• The Hebgen Dam, an earth dam located in Montana only a few hundred yards from the fault of the Hebgen Lake earthquake of 1959 (*M* 7.1), survived the shaking without failing.

The adequate performance of these structures under strong shaking gave the profession a high level of confidence in the design procedures in use at the time.

In the 1960s and early 1970s a number of events occurred that caused engineers to reevaluate the adequacy of this approach:

• The observation was made that the pseudostatic method of analysis would not adequately predict the slope failures that occurred at many places in Alaska in the 1964 Alaska earthquake (*M* 8.3).

• Major cracking occurred in the Koyna Dam, in India, a 340-foot-high concrete gravity dam, subjected to a near-field *M* 6.4 earthquake in 1967.

• Major cracking developed near the top of the Hsingfengkiang Dam, a 345-foot-high concrete buttress dam in China, as a result of a near-field *M* 6.1 earthquake in 1962.

• A near failure occurred at the Lower Van Norman (San Fernando) Dam, and significant sliding occurred in the Upper Van Norman (San Fernando) Dam as a result of the San Fernando earthquake of 1971. In spite of the fact that both of these dams were judged to be adequately safe on the basis of pseudostatic analyses made before the earthquake, their performance cast doubt on the validity of this approach.

• Accelograph records showed peak accelerations during earthquake shaking greater than 0.3 g.

These events, and others, led to an increasing concern that the pseudostatic method of analysis could not always predict the safety of dams against earthquake shaking. At about the same time, new tools for making improved analyses of seismic response had become available (finite element methods and high-speed computers), and investigations were made of the ability of dynamic response analyses to provide better insights into probable field performance. The results of these studies were extremely encouraging and sufficiently convincing that by 1973 three major changes occurred.

1. In the late 1960s the California Department of Water Resources adopted methods of dynamic analysis for design of the State Water Project.

2. The Division of Safety of Dams of the California Department of Water Resources decided to require owners of many earth dams to reevaluate the seismic stability of the dams using dynamic analysis methods, regardless of the results indicated by pseudostatic methods of analyses.

3. The Earthquake Committee of the International Commission on Large Dams (ICOLD, 1975) recommended that:

As regards low dams in remote areas, they may be designed by the conventional (pseudostatic) method for any type of dam. As regards high gravity or arch dams or embankment dams whose failure may cause loss-of-life or major damage, they should be designed by the conventional method at first and they should then be subjected to dynamic analyses in order to investigate the deficiencies which might be involved in the seismic design of the dam in case the conventional method is used.

These requirements and recommendations led to a major review of design procedures by many design agencies, as a result of which increasing emphasis was given to the use of dynamic analysis methods where appropriate. Thus, among federal agencies in the United States, methods of analysis currently (1984) in use are generally as follows.

Concrete Dams

• defensive design measures (studies to ensure foundation and abutment integrity, good geometrical configuration, effective quality control, etc.);
• pseudostatic analysis methods, commonly used to check sliding and overturning stability for gravity or buttress dams and in areas of low seismicity; and
• dynamic analysis methods (assuming the dam to consist of linear elastic, homogeneous, isotropic material) to determine the structural response and induced stresses in areas of significant seismicity.

Earth Dams

• defensive design measures (ample freeboard, wide transition zones, etc.);
• for reasonably well-built dams on stable foundation soils, no analysis required if peak ground accelerations are less than about $0.2\,g$;
• a dynamic deformation analysis for dams constructed of or on soils which do not lose strength as a result of earthquake shaking and located in areas where peak ground accelerations may exceed about $0.2\,g$; and
• a dynamic analysis for liquefaction potential or strain potential for dams involving embankment or foundation soils that may lose a significant portion of their strengths under the effects of earthquake shaking.

Private engineering companies follow generally similar practice. These procedures necessarily require an evaluation of the seismic exposure of the dam and, in cases where dynamic analysis procedures must be used, a determination of the earthquake motions (either in the form of a response spectrum or in the form of time histories of accelerations) for which the safety of the dam must be evaluated and for which the safety of the dam must be assured. The selection of these motions involves considerations of seismic geology, seismicity, probability of occurrence, consequences of failure or damage, and public safety.

5
Design Flood Estimates: Methods and Critique

TYPES OF APPROACHES

As indicated in Chapter 3, most current criteria for spillway adequacy can be placed in two general classes: (1) criteria based on probable maximum precipitation (PMP) estimates or probable maximum flood (PMF) estimates derived from PMP estimates and (2) criteria based on either floods or rainfalls having specified probabilities or average return periods. A third approach to sizing spillways by analysis of risks to downstream areas from dam failures has come into use, and a variation of this approach, involving economic analyses of risks and costs of prevention, is also in use. The PMF is considered to define the upper range of flood potential at a site. Because of the methods used in developing safety evaluation flood estimates, the criteria based on PMP and PMF estimates are termed the deterministic approach, while those based on rainfalls or floods of specified frequencies are labeled the probabilistic approach. These approaches to establishing appropriate design flood estimates are discussed in this chapter, and further details relevant to the deterministic and probabilistic methods are presented in Appendixes C and D. Details of the risk analysis approach, along with examples of its use, are presented in Appendix E.

DETERMINISTIC APPROACH

Summary

In estimating a PMF or any design flood by deterministic methods, several tools of meteorology, hydrology, and hydrologic engineering are employed

44

to synthesize a hydrograph of inflow into the reservoir and to model or simulate the movement of the flood through the reservoir and past the dam. The various steps in such analysis, all of which have been discussed in the NRC's report, *Safety of Existing Dams: Evaluation and Improvement,* Chapter 4 (1983), are generally as follows:

1. dividing drainage area into subareas, if necessary;
2. deriving runoff model;
3. determining PMP using criteria contained in NOAA Hydrometeorological Report series;
4. arranging PMP increments into logical storm rainfall pattern;
5. estimating for each time interval the losses from rainfall due to such actions as surface detention and infiltration within the watershed;
6. deducting losses from rainfall to estimate rainfall excess values for each time interval;
7. applying rainfall excess values to a runoff model of each subarea of the basin;
8. adding to storm runoff hydrograph allowances for base flow of stream, runoff from prior storms, etc., to obtain the synthesized flood hydrograph for each subarea;
9. routing of flood for each subarea to point of interest;
10. routing the inflow through the reservoir storage, outlets, and spillways to obtain estimates of storage elevations, discharges at the dam, tailwater elevations, etc., that describe the passage of the flood through the reservoir. (This is essentially a process of accounting for volumes of water in inflow, storage, and outflow through the flood period. If there are several reservoirs in the watershed, the reservoir routing is repeated from the uppermost to the most downstream reservoir, in turn.)

If the routing shows that the dam would be overtopped and if it is assumed such overtopping will cause the dam to fail, the resulting flood wave may be routed through the downstream valley to give a basis for assessment of damages.

Probable Maximum Precipitation (PMP) Evaluation

Of the factors that have an influence on the magnitude of the probable maximum flood (PMF), the intensity and duration of rainfall are the most important; hence, considerable discussion of such rainfall estimates follows here and in Appendix C. The rainfall that produces the PMF is termed the probable maximum precipitation (PMP). Its definition (which goes back to the early 1950s) is "the theoretically greatest depth of precipitation for a given duration that is physically possible over a given size storm area at a particular geographical location at a certain time of the year." Several other definitions have been given for PMP, but, in any case, PMP should be termed an estimate, as there are yet unknowns and unmeasured atmospheric parameters that are important to extreme rain storms.

In the literature prior to about 1950, the term maximum possible precipitation (MPP) was used for estimates of most extreme rainfall. This was changed to probable maximum precipitation because the latter term implies a somewhat less extreme and absolute estimate and reflects the uncertainties involved in estimating maximum precipitation potentials. In practice, both MPP and PMP have essentially the same meaning.

PMP estimates usually greatly exceed the rainfall amounts most people experience. Table 5-1 lists current PMP estimates for the Washington, D.C., area as read from charts developed by Schreiner and Riedel (1978). By contrast, the maximum 24-hour *point* rainfall that has been recorded in 114 years of record at Washington, D.C., is 7.13 inches. However, within 125 miles, near Tyro, Virginia, 27 inches of rain fell in 12 hours during a storm occurring August 19-20, 1969.

In Chapter 4 a brief history is given of the development of the concepts involved in PMP determinations. Such determinations are described in some detail in Appendix C and summarized sequentially as follows:

1. Study major rain storms to determine maximum areal rainfalls and ascertain, as well as possible, the meteorological factors important to the rainfall.

2. Transpose the major storms within topographically and meteorologically homogeneous regions. (Adjust the storm rainfall by multiplying it by the ratio of an index of maximum atmospheric moisture in transposed location to that where the storm occurred.)

TABLE 5-1 PMP Estimates (to Nearest 0.5 Inch) for Vicinity of Washington, D.C.

Area	Time Duration (h)				
(sq mi)	6	12	24	48	72
10	27.5	32.5	36	40	42
200	19.5	23	27	31	32
1,000	14	18	22	25	26
5,000	8.5	12	15	19	19.5
10,000	6.5	9.5	12.5	15.5	17

SOURCE: Schreiner and Riedel (1978).

3. Adjust the rainfall (for each transposed storm) by the ratio of maximum atmospheric moisture in place of occurrence to that which existed during the storm.

4. Smoothly envelop the resulting rainfall values durationally, areally, and if generalized PMP is being developed, regionally. Explanations should be given for discontinuities.

The resulting rainfall values may be regarded as PMP if sufficient storms were available to justify the assumption that the maximum rainfall potential has, in fact, been assessed. Appendix C describes some procedures that give the user a feel for how conservative the PMP is. One shows the ratios of PMP depths for a 10-square-mile area and 24-hour duration to the point rainfall depths for 100-year average return period and 24-hour duration at the same locations. These ratios range between 2 and 6 in the United States. Another study shows that 90 known experienced storm rainfall depths are equal to or greater than 50 percent of PMP, both rainfall categories being for 24 hours and 10 square miles. Many more storms exceed 50 percent PMP for several combinations of durations and area sizes (Riedel and Schreiner, 1980). Figure 5-1 shows a comparison of generalized PMP for 200 square miles and 24 hours determined in 1947 with that in 1978 for the same region (United States east of the 105th meridian). This comparison shows a general increase in PMP estimates over the 31-year period. About 65 percent of the region shows an increase between 10 and 30 percent. From such comparisons it can be concluded that the PMP estimates for the United States from generalized charts determined by the National Weather Service are most likely on the low side when evaluated with reference to the accepted definition of PMP and that future increases in PMP estimates are probable. However, such future increases in estimates are likely to be incrementally less, in general, than those of the past during the period that the science of hydrometeorology was rapidly developing and the data base on past extreme events was being accumulated. Also, studies to redetermine probable maximum precipitation estimates for areas in the Tennessee Valley made in 1978 resulted in some lower estimates than those developed for the same areas in 1968; then more intensive studies sometimes can result in lower estimates even if more data became available.

PMP estimates developed as indicated have been widely (but not universally) accepted as the appropriate basis for design of spillways for large dams where failure of the structure by overtopping cannot be tolerated. The increases noted in PMP have posed difficult questions as to what should be done with spillways at existing dams or those already under construction, where the spillways were adequate under previous criteria but would not be adequate with the revised PMP estimates.

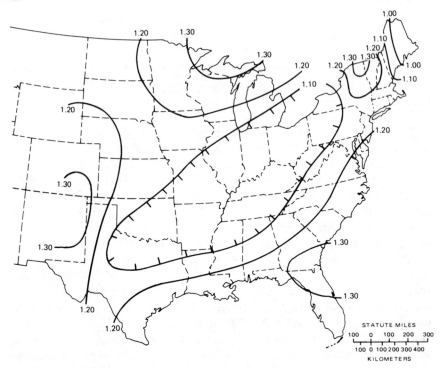

FIGURE 5-1 Comparison of generalized PMP estimates for 24 hours and 200 square miles made in 1978 with those made in 1947. (Ratios shown are 1978 values/1947 values.) Sources: The 1978 data are from Schreiner and Riedel (1978). The 1947 data are from U.S. Weather Bureau (1947).

Antecedent Conditions

An important consideration in assessing the impact of any extreme flood on a project is the spectrum of antecedent conditions. These conditions include soil moisture, snowpack and water content where applicable, expected reservoir levels, state of vegetation, intended use or uses of the project, probability of preceding and subsequent precipitation events, and ambient temperatures.

It is generally recognized that even though a hypothetical flood, such as the PMF, is an extreme event to be adopted as a basis for design, it should be conceived in hydrologic and meteorological reasonableness. The antecedent conditions, such as expected reservoir levels, existing snowpack, and soil moisture, are considered in the context of the causative event of the primary flood. For example, if the extreme floods for moderate to large basins result from tropical storms that occur from late summer through fall and highest reservoir levels always result from spring snowmelt floods, the two events are not usually combined. In another instance, the test of reasonableness might indicate that in mountainous country, major floods for small basins could result from a heavy rain producing convective thunderstorms, but not of the PMP magnitude, that could occur concurrently with the seasonal snowmelt-generated peak runoff. Generally, it can be considered meteorologically unlikely for the *maximum* snowmelt runoff flood to occur in combination with any extreme precipitation event. In each situation, it is desirable to evaluate meteorologic reasonableness of the criteria rather than to apply arbitrary rules.

Soil moisture and the state of the vegetation affect loss rates and basin runoff characteristics. These vary seasonally and should be evaluated to determine appropriate conditions for occurrence prior to the PMF. Antecedent (and subsequent) storm conditions can have an impact on the adequacy of any reservoir design.

The effects of the antecedent (or subsequent) storm are related to storm magnitude, area covered, season of occurrence, dry interval between events, and geographic region of the primary rainfall event. It is not feasible to establish general criteria for such conditions. Each situation should be carefully evaluated to assure that assumed conditions are sufficiently conservative but not atypical of the region.

The basic purpose for which the project is designed governs the rules for reservoir operations. Customarily this rule curve is the end-product of study of annual inflow patterns, the schedule of need for releases, the use of these releases, and any site-specific restrictions imposed on reservoir discharges and reservoir levels by downstream channel capacities and riparian interests. These are additional items that must be weighed in importance, and from this analysis is obtained a reservoir level to be used as the initial point in the routing of the design flood and the antecedent storm through the storage and spillway facilities of the site. Usually to avoid increasing downstream flood damages, any assumed reservoir operation plan should include the requirement that no operating procedure will increase the peak reservoir outflow over the peak natural inflow unless specific flood easements to accommodate excess flow downstream have been obtained. As an example, drawdown of reservoir storage in anticipation of a tropical event, which ultimately tracks away from the project catchment, could produce downstream flows in excess of the inflow that did occur. If this excess flow exceeded channel capacity, it could lead to downstream property damage and claims. Often, however, the orderly drawdown of reservoir storage in those climates where the annual flood event is a snowmelt case, is an accepted practice. This acceptance is based on the fact that the seasonal snowmelt runoff volume may be closely estimated with present-day prediction techniques.

Reservoir Routing

The ability of any reservoir to release water downstream is determined by the project components. Usually these components are comprised of some combination of spillway gates, flashboards, stanchion bays, low-level outlets, and turbines if the project generates power. In preparation for routing the design flood through the project storage and these facilities, rating curves of discharge versus head must be computed for each applicable component and for the proposed combination of discharge facilities to be used.

As a case in point, generally a project designed for flood control will limit releases to the bankful channel capacity downstream up to the point where the inflow has occupied the allocated flood storage volume. From that point on, the typical emergency spillway operates so that the discharge is a function of the head on the spillway, the incremental flood storage usually being fairly small.

In those cases where flood control is not the primary purpose of a project, a gated spillway is usually operated when the inflows exceed the capacity of outlet works, including hydropower turbines. This gate operation is usually carried out so that the reservoir level remains constant, thus the outflow is matched with the inflow up to the point of maximum gate operation. Some regulatory bodies require a gate capacity such that a specific percentage of the PMF can be accommodated with one gate inoperable without the dam being overtopped.

The routing of the PMF inflow hydrograph through the available reservoir storage and the spillway facilities of the project utilizes some variation of the volumetric conservation equation:

$$I - O = \Delta S$$

where

$$I = \text{reservoir inflow}$$
$$O = \text{the outflow or discharge}$$
$$\Delta S = \text{change of storage in reservoir}$$

The mechanics of applying and solving this basic equation are given in the standard hydrologic texts and will not be described herein.

The preceding discussion is a simplified explanation of what can be very complex operational studies of effect of extreme storm rainfall on a basin and a reservoir. In a typical safety evaluation, judgment is required on many factors. Experienced hydrologic engineers may differ to such an extent that their assessments of project safety will be affected. Some may tend to adopt the most critical value for each parameter involved and, thus, tend to make the PMF estimate and routing excessively improbable. Others may tend to regard each of the aspects as independent of the PMP and other factors and adopt average or medium values for each factor. Detailed studies of each situation may provide a guide to selection of the most logical array of choices for that project, but it is not practicable to attempt to formulate rules for such choices at all projects.

THE PROBABILISTIC APPROACH

As indicated in Chapter 3, safety evaluation floods for small dams, where no serious hazards would exist downstream in the event of breaching, are usually based on rainfall-runoff probability estimates. As discussed in Chapter 4, the occurrence of floods in some basins far larger than could have been predicted by probability studies of prior stream records has discouraged the use of probabilistic methods for estimating extremely rare floods. However, the state of California uses estimates of floods with average return periods of 1,000 years as minimum floods for evaluating safety of low-hazard dams. In contrast to practices in the United States, the criteria of the Institution of Civil Engineers call for use of estimates of floods with average return periods of 10,000 years in the British Isles. Past experience has indicated that estimates of magnitudes of very rare floods developed by probabilistic methods are even more likely to change as additional basic data become available than flood estimates developed by deterministic methods discussed earlier in this chapter.

Some of the basic principles of probability studies are discussed in Appendix D. Flood-frequency analyses as discussed in Appendix D produce estimated instantaneous peak flows with no estimate of flood volume. A complete hydrograph is needed to perform a flood routing through the reservoir of a given dam. Such a flood hydrograph can be synthesized by utilizing an observed hydrograph from a major historical flood and increasing the ordinates of that hydrograph by the ratio of the peak flow determined by frequency analysis to the observed peak flow.

Rainfall frequency data can provide another way of synthesizing a flood hydrograph with desired estimated frequency of occurrence. Such a hydrograph can be developed utilizing an appropriate rainfall-runoff model and rainfall frequency data such as available from NOAA *Atlas 2* (Miller et al., 1973).

Once a hydrograph of inflow into the reservoir, representing the safety evaluation flood, is obtained, the routing and analyses procedures are essentially as described for the deterministic approach.

THE RISK ANALYSIS APPROACH

Methods of evaluating dam safety by analysis of effects of hypothetical dam failures on downstream areas and on project benefits and costs have been advanced in recent years. These methods do not depend on adoption in advance of specific bases or criteria for dam safety evaluation floods but depend upon site-specific analyses to select the flood appropriate to the safety evaluation. Two types of analysis are in use: (1) those that evaluate only the hydraulic effects of dam failures and (2) those that go further and make an economic analysis to determine the design that has minimum total cost. Appendix E discusses and gives an example of the latter approach.

Through computerized modeling of floods and dam break inundation mapping, the safety evaluation flood can be selected at the flood peak level where downstream flood damages would not be increased by the overtopping of the dam. In other words, through an iterative trial-and-error computation process, the spillway is sized so that all significant downstream flood damages, from spillway releases and other sources, will have occurred before the dam fails by overtopping. This approach is allowed, as an alternative to selecting a design flood from a generalized chart, by the U.S. Bureau of Reclamation and by several state dam safety programs including Arizona, Colorado, Georgia, North Carolina, Pennsylvania, and South Carolina. Pacific Gas and Electric Company, a private utility firm, also uses this approach to evaluate existing dams. This alternative is also proposed in draft guidelines prepared by a working group of ICODS, the Interagency Committee on Dam Safety (1983). One of the limitations of this general approach of evaluating only hydraulic effects is the possibility that subsequent downstream development will encroach on the dam-break inundation area and thus change the conditions which determine that dam failure would cause no further significant damages. Another limitation is that potential damage to project structures and the value of project services (e.g., water supply) are not reflected in the analysis.

A more complex version of this approach is to estimate the dollar cost of damage, loss of services provided by the dam, and construction costs of several design alternatives; to estimate the probability of failure for each alternative; and to select the final design at the lowest risk-cost combination. No dollar value is assigned to human lives; in some cases, downstream warning systems and evacuation plans have been relied on to avoid putting human lives into the risk-cost analysis.

The quantitative risk-cost analysis approach has been applied to very few dams and is such a recent development that it can barely be called "current practice." However, its use in selecting design standards can be expected to increase in coming years. Risk analysis procedures have been defined by the Bureau of Reclamation (1981a) for internal use. The Interagency Committee on Dam Safety (1983) encourages site-specific breach-routing studies as part of the hazard assessment for proposed dams and suggests that risk-based analysis may be a basis for decision on selection of the safety evaluation flood at particular existing dams.

CRITIQUE

Some of the deficiencies or limitations observed in currently used criteria and procedures relating to provisions for safety of dams from extreme floods are inherent to any attempt to deal with the random-chance nature of rare floods. This discussion is not intended necessarily as criticism of the criteria or procedures nor of those groups who use them. The comments herein are based on the array of criteria and practices in current use and may not always be applicable to the programs of the Corps of Engineers and the Bureau of Reclamation.

The goal of dam safety is to limit the risks from dam failures to acceptable levels. Probability of failure is controlled partly by design standards and partly by quality of design, construction, inspection, operation, and maintenance. Ideally, hazard, failure probability, and acceptable damage would be quantified for the site-specific conditions of each individual existing or proposed dam in order to establish site-specific standards for achieving this goal. With few exceptions, current practices do not involve quantification of these three critical elements for each dam.

Instead, the most widespread current practice is to classify dams in three broad, not well-defined, qualitative damage potential categories (i.e., high, intermediate, and low hazard) and to somewhat arbitrarily assign one of three or four grades or ranges of design standards to each dam depending on its height, storage capacity, and qualitative hazard rating. Current practice treats all of the elements needed for selecting design standards in a generalized way; thus, the appropriateness of the design standards as applied to individual dams is generally unknown.

In defense of this current general practice, it must be recognized that most of the scores of federal and state regulatory agencies each have hundreds to thousands of dams under their jurisdictions. Given their limited resources, as a practical matter, they must use a generalized system of assigning design standards according to generalized hazard and size classifications, at least as an interim step until more detailed site-specific studies can be made. However, the wide range of hazard versus size versus design standards among the various agencies (see Table 3-3 and Appendix A) reflects a lack of uniformity even within the generalized current practice.

This lack of uniformity in dam classification and safety design standards appears to result from three main factors: (1) lack of interagency and inter-governmental communication, (2) variations in engineering judgment in selecting the generalized standards, and (3) variations in public policy atti-tudes at the times the standards were selected. In any case, a critique of present practices must point out that, though a generalized approach to selecting design standards is justified as a practical interim step, there is a need for more uniformity among the various federal and state agencies in establishing size and hazard definitions and correlative design standards.

Dam Classification Systems

Even if we recognize the need for generalized hazard versus size versus design criteria classifications, the almost universally used high-, intermedi-ate-, and low-hazard classes are not well defined. Qualitative definitions for these terms, such as those used by the Soil Conservation Service and Corps of Engineers, are followed by most federal and state agencies.

Some examples of the lack of uniformity in defining "hazard" among the many regulatory agencies are as follows:

• One federal agency estimates damage potential (i.e., hazard class) as-suming "sunny day" failure, while another federal agency assumes failure only during "floods" as a basis for its hazard classification.

• In defining high-hazard dams, agencies use such terms as "probable" loss of life, "possible" loss of life, "rural" and "urban" houses downstream, and one agency says that more than 10 houses in a dam-break floodwave places the dam in a high-hazard category. One state agency says a dam is high hazard if there is potential "extensive" loss of life, significant hazard if the dam would endanger "few" lives; another defines a dam as high hazard if there would be "substantial" loss of life, intermediate hazard if a "few" lives would be lost. In contrast, some agencies consider the probable loss of one human life as a high-hazard condition.

• No agencies define the dollar value of "extensive," "significant," or "minor" economic losses in their hazard classes.

An attempt to quantify these hazard definitions was made by the North Carolina Dam Safety Program in 1980 and 1982. Questionnaires asking respondents to quantify the minimum values for each hazard class were distributed among the program staff and among participants at a Southeast-ern Regional Dam Safety Conference. The results (heretofore unpublished)

TABLE 5-2   Boundaries for Hazard Classes

Hazard Classification	Mean Values of Opinions	
	Probable Loss of Life	Economic Loss[a]
Low	0	0 to $30,000
Significant	0	$30,000 to $200,000
High	1 or more[b]	Greater than $200,000

[a]Includes downstream damages, but not cost of dam or value of services provided by reservoir.
[b]Strong consensus that loss of one life defines high hazard.

reflect an extremely wide range of opinions but indicate the following median opinions from the 46 individual respondents for quantifying the boundaries on hazard classes (Table 5-2).

Undoubtedly, others will disagree with these evaluations, but such an effort toward more specific hazard definitions could be a step toward a more uniform approach to setting generalized standards.

Another weakness in current practice is that, generally, downstream hazards are only roughly estimated through judgment based only on visual inspection. This practice is a reflection of limited resources rather than technology, however, and it is reasonable to believe that essentially all practitioners recognize the desirability of inundation mapping through breach-routing methods.

Spillway Capacity Criteria

As shown in Chapter 3, there has been general agreement, with some exceptions, that the spillways of large, high-hazard dams should be able to pass the probable maximum flood without the dam being overtopped. All federal agencies agree with this standard. Only a few states indicated that smaller floods are used as criteria for spillway capacity at such dams. One other type of exception sometimes encountered involves concrete dams on solid rock foundations. Indicated overtopping of such a dam during the probable maximum flood may be permitted by some agencies, if the rock at the toe of dam is judged able to withstand the hydraulic forces imposed and the stability of the dam would not be compromised otherwise. For smaller dams and those with lower hazard ratings, there is much greater divergence in views concerning appropriate spillway capacity requirements. As noted in Chapter 4, the rationale for some of the spillway capacity criteria in use seems questionable, particularly the criteria based on an arbitrary percentage of the probable maximum flood or an arbitrary percentage of a flood of specified probability or criteria that combine the probabilistic and deterministic approaches. The problem with such a criterion, based on an arbitrary percentage of a derived flood or on arbitrary combination of floods developed from differing concepts, is that it permits no direct evaluation of the relative degree of safety provided.

While regional differences in climate, geography, development, etc., could justify some of the differences in spillway capacity criteria, it appears that not all the criteria could be efficient in limiting risks of dam failures to acceptable limits or in protecting the public interest. Efforts to secure more uniform approaches to specifying spillway capacity should be encouraged, but such effort is considered beyond the scope of this report. The newly established Association of State Dam Safety Officials may wish to consider action toward such a goal.

Some differences among agencies have been noted in practices followed in developing probable maximum flood estimates from probable maximum precipitation values. These differences relate to assumptions regarding antecedent rainfalls, initial reservoir levels, arrangement of precipitation values, runoff models, etc.

The committee has found general agreement in the following observations regarding current spillway capacity criteria:

• Interpretations of data from past storms and storm model concepts are required to make estimates of PMP.

• As shown by past experience, PMP estimates can change as more data become available; thus, the PMP estimate cannot be regarded as a fixed criterion, but confidence in the estimates should rise with successive PMP estimates for a given locality.

• The probability that rainfall will equal or exceed current PMP estimates is indeterminate but probably not uniform for projects in different parts of the country.

• In order that judgments can be made on appropriate allocation of resources, it would be desirable to be able to express spillway design flood criteria in terms of annual probabilities.

• As has been found previously, statistical studies of data from past floods may serve to indicate the minimum spillway capacities that should be considered but generally do not provide reliable basis for spillway design if there are significant or high hazards downstream in the event of dam failure.

• As a dam owner or as a regulator of dams in the interest of public safety, a government agency should seek to achieve a proper balance between costs to improve dam safety and risks to the public.

• It is appropriate that dam safety criteria recognize differences in consequences of failure.

• It is good public policy to require management plans (such as warning systems, evacuation plans, and operating rules) to reduce hazards (principally to lives) of dam failures, but the long-term use of such plans should not be a substitute for work to remedy serious safety deficiencies. (The committee notes that maintenance over long time periods of an effective emergency management system to avert loss of life from failure of a dam would require continuing dedicated efforts and support to maintain and operate hardware, to train and inform emergency action personnel and the public, to establish and maintain institutional arrangements, and to upgrade the system to meet changes in the area to be served.)

• Even though some problems with current PMP estimates have been noted, such estimates still offer the bases on which the engineering profession has the most confidence for sizing spillways of new, large dams in the high-hazard category.

• Each existing large, high-hazard dam having a spillway that fails to meet current PMF criteria should be considered separately. It does not seem appropriate to adopt fixed rules for such situations. Each study should consider how deficient the project is under current criteria and the relationship of the allocated spillway capacity to other flood criteria. If the deficiency relates to change in safety evaluation criteria (such as an increase in PMP estimates), the reasons for such change and their relationship to the project in question should be critically examined.

Risk-Cost Analyses

As described earlier in this chapter, the risk analysis approach has provided a significant trend toward improved assessments and toward selecting more rational, site-specific spillway evaluation standards within the last few years. Though risk-cost analyses may appear to represent the most desirable approach to the goal of dam safety (i.e., in quantifying hazard, failure probability, and acceptable damage) at this time, this method has certain important problem areas or limitations that the user needs to consider. An ICODS critique of the risk-cost analysis method mentions the following points.

• Estimates of the probability of exceedance of extreme hydrologic events are imprecise, whereas the total costs associated with different alternatives may be sensitive to these estimates.

• Those factors which cannot be measured in economic terms such as loss of human life, social losses, and environmental impacts are more difficult to reflect in the risk analysis but may be the most important in making decisions.

• Results of a quantitative risk-based analysis reflect probable annual costs but a dam failure may result in a single catastrophic loss from which the owner and many others may not recover; thus the relevancy of the analysis to the interests of the parties involved may be questionable.

Additional problem areas noted in risk-cost analyses are as follows:

• Future development below a dam is usually unpredictable and may invalidate the risk-cost determination and the safety evaluation flood selected for present (or inaccurately predicted future) conditions.

• There may be a tendency to rely on downstream warning systems to eliminate loss of human lives from the analyses and thus to determine the lowest risk-cost design standard. However, the real effectiveness of a downstream warning system may be questionable and reliance on such systems may give a false sense of security to design engineers, dam owners, and residents below a dam.

• Risk analysis places heavy emphasis on hydraulic evaluations of unusual situations at dams and downstream. However, the reliability of flood- and dam-break-routing models has not been sufficiently determined; the models have been checked against only a few actual dam-break floods. Yet the accuracy of modeling flood and dam-break inundation areas is often critical in a risk-based analysis. One unknown in even the best dam-break-routing models is the "rate of breach development" to assign during the modeled failure; the estimate of damages from relatively small reservoir dam failures is often extremely sensitive to rate of failure (rate of reservoir release), particularly for "sunny day" failure conditions. Also, in quantifying downstream damage predictions (hazard) in risk-based analyses, the water depth/velocity and debris load required to damage various structures are sometimes uncertain. This can have a critical effect on the accuracy of the computed lowest risk-cost.

• The depth and duration of overtopping that various dams can withstand without failure are unknown. It may be desirable economically, and physically safe, to allow some overtopping of existing dams in a risk-based analysis, but there are no reliable data on tolerable limits. This has dramatic economic implications nationwide that will not be resolved by risk-based analyses.

• Although a risk-based analysis may not be expensive if compared to probable costs of remedial measures to improve dam safety, for some dam owners such an analysis may not appear to be economically feasible for smaller (though high- or intermediate-hazard) dams. Further, some regulatory agencies may not have the financial and technical resources to conduct risk-based analyses on tens of thousands of dams in order to set appropriate site-specific design standards.

The above-discussed areas of potential problems in application of risk analysis technique do not, necessarily, detract from the usefulness of this type approach, as long as they are recognized and provided for. In fact, it is expected that, with more experience and research, these limitations may be minimized. Other factors that make risk-based analysis an *attractive* technique are as follows:

• Risk-based analyses, as presently performed, generally are not intended to replace appropriately conservative design standards. Rather, risk-based analyses provide additional information to decision makers to help them decide how limited funds can best be allocated to reduce risks.

• Risk-based analyses are not intended to provide a sole basis for making decisions. They only provide a portion of the information needed.

• By performing sensitivity studies, many of the problems with performing a risk-based analysis can be minimized and the results bounded.

• The process of performing a risk-based analysis often uncovers factors or sensitivity relationships that might otherwise not be identified.

• Those factors that cannot be measured in economic terms, such as loss of human life, can be accounted for in separate risk-based analyses and given the appropriate weight (as implied in Appendix E).

Overview

This critique of current practices has focused on three levels of sophistication in setting standards: (1) the widespread generalized approach, relying largely on judgment to assess hazard and selecting design standards based on loosely defined categories; (2) using site-specific dam-break-routing studies to better define hazard and to select a spillway design flood without quantifying risks and costs; and (3) risk-based analysis, which extends the second category by attempting to quantify all of the significant variables in selecting standards. Some of the main strengths and deficiencies associated with each of these three levels have been discussed. Two other deficiencies must be pointed out in a broad overview of current practice. First, about one-half of the states either have *no standards* for nonfederal dams or have seriously inadequate implementation of standards (Tschantz, 1983, 1984). Since there are well over 60,000 nonfederal dams in this country, this current practice (really, lack of applying any standards) has serious national implications regarding the achievement of safety goals for dams. Second, most standard-setting efforts have been focused (as does this report) on large, high-hazard, federally owned dams where it is clear that very high standards must be applied and where public investment in the dams and their services is very important. However, smaller nonfederal dams (U.S. Army Corps of Engineers, 1982) pose the greatest aggregate of risks nationally, and their wide range of sizes and hazards requires a wide range of design standards. Focus on the various aspects of the PMP or PMF for large, high-hazard dams has tended to detract from the need for developing appropriate standards for tens of thousands of smaller, yet very important dams.

In summary, design standards based on size ranges and general hazard classifications of dams are a necessary evil from a regulatory or administrative standpoint, but the diverse and ambiguous definitions within this predominant current practice reflect a serious lack of uniformity among federal and state agencies in applying this approach. The generalized hazard definitions are vague, and the appropriateness of the design standards applied to the size/hazard ranges is generally unknown, leaving the appropriateness of the generalized standards at specific dams even more in doubt than they reasonably could be. Exceptions exist for very large, very high-hazard dams for which there is a clear consensus that something like the probable maximum flood is the appropriate inflow design flood or is most likely the best of the available alternatives for such applications. However, this design flood is not necessarily appropriate for the thousands of existing smaller (yet high-hazard) dams. Where site-specific studies are economically feasible, selection of the design standard for each dam through dam-break routing studies without placing a dollar value on predicted damages is definitely a temporary improvement over the generalized standards but is limited by the possibility of future downstream development. Finally, risk-based analysis, which attempts to quantitatively balance the total cost of alternative design standards against probability of failure, has its own limitations in its present state of development, but can provide information useful to decisions involving making dams safe from extreme floods.

6
Design Earthquake Estimates: Methods and Critique

The occurrence of an earthquake is a physical process which, in principle, is completely understandable and, if enough data were available, would be predictable. Strains and stresses are being built up in certain regions of the earth's crust, and when the strength of the material is exceeded, a stress failure occurs. The sudden release of stress that is triggered by the failure generates stress waves that propagate in all directions and produce earthquake shaking at the surface of the ground. The stress failures that produce destructive shaking are initiated at depths of a few miles or a few tens of miles, and at these depths the weight of the superposed rock produces large compressive stresses and, as a result, only shearing stress failures can occur.

Over the past millions of years many stress failures have occurred with relative displacement across the failure surface, and these surfaces can be identified by geologists when seen on the surface of the ground and at depth by geophysical prospecting methods. Geologists have named these old stress failure surfaces "faults." Such faults are surfaces of weaknesses, and present-day stress failures invariably occur on existing faults, such as those shown in Figure 6-1 for the state of California. Thus, earthquakes could be predicted if we had knowledge of the locations and geometry of faults, the existing stress distribution over the surface of the fault, the strain rates in the earth's crust, the value of the failing stress on the fault, and the requisite physical properties of the rock in the region of the fault being studied. However, because of the difficulty of obtaining the necessary data, such information is not sufficiently well known to make a scientific determination of the location, time of occurrence, and magnitude of earthquakes.

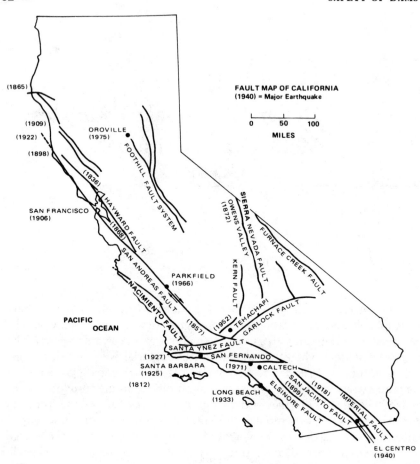

FIGURE 6-1   This diagram shows the prominent faults in California. The maximum capable earthquake on a fault is sometimes taken to be that event generated by slip traversing three-fourths of the length of the fault. Thus, great earthquakes can be expected on large faults. Small earthquakes can be expected on both long and short faults. The dates within parentheses indicate the locations of major earthquakes.

An additional difficulty in estimating the nature of ground shaking is that as the seismic waves travel away from the fault, they traverse heterogeneous earth and are affected by reflections and refractions at the heterogeneities. Therefore, the shaking at a point on the surface of the ground depends not only on the details of the source mechanism but also on the details of the travel path, neither of which are well known. To circumvent this lack of knowledge, data have been collected on historical earthquakes, including location, date, magnitude, intensity, etc. In addition, data are collected on the prehistory of earthquakes, including identification of faults, estimates of most recent fault displacements, and crustal plate movements, which can throw light on seismic activity. The historical data and the prehistorical data (over geologic time) provide the bases for estimations of seismic hazard.

At present, to estimate seismic hazard, either statistical analyses of motion characteristics must be used, or near upper bounds must be specified. If the magnitude is taken to be that of the largest possible earthquake that can be expected to occur along the fault, the event is called the maximum credible earthquake (MCE). The motion at the dam site resulting from such an earthquake is called the maximum credible earthquake motion, or sometimes simply, maximum credible earthquake. For example, along the southern portion of the San Andreas fault in California the average return period for earthquakes of magnitude 8-plus is estimated to be approximately 150 years.

At places where the historical record of earthquakes is short in comparison with the recurrence time of the MCE, the MCE may be larger than the largest historical earthquake. For such cases, different investigators employ different empirical relations to estimate the magnitude of the MCE. These methods usually take into account, either objectively or subjectively, the notion of a "reasonable" return period based upon the present tectonic regime; that is, the MCE is not taken to be the earthquake that will not be exceeded in some extremely long period of time, such as 100 million years.

When adequate information is available, deterministic methods are used for estimating design earthquake motion for dams when loss of a reservoir would result in loss of human life and/or substantial economic loss, and these methods are used most often for other critical facilities whose catastrophic failure would produce similar kinds of losses. But, increasing attention is being devoted to the application of probabilistic-risk analysis methods for earthquake-resistant design criteria for nuclear reactor facilities. Such methods are also used, in some cases, to provide background information on seismic hazards of major dams.

DETERMINISTIC-STATISTICAL METHOD

The deterministic-statistical method requires certain basic information: earthquake magnitude, smallest distance from the fault or the earthquake source zone to the dam site, equations or curves relating magnitude and distance to peak ground acceleration, peak ground velocity and duration of strong ground shaking, and sometimes a site correction for the soil layer above the bedrock at the dam site.

Uncertainty is associated with each phase or step of the deterministic estimation of strong ground motion at the site. Empirical equations or curves, such as shown in Figure 6-2, that relate fault rupture length to earthquake magnitude often are used for estimating the MCE. However, there is appreciable scatter in the data that are used to determine the fault rupture length versus magnitude relation, because of variations in some of the other physical characteristics of the earthquake source. Thus, a statisti-

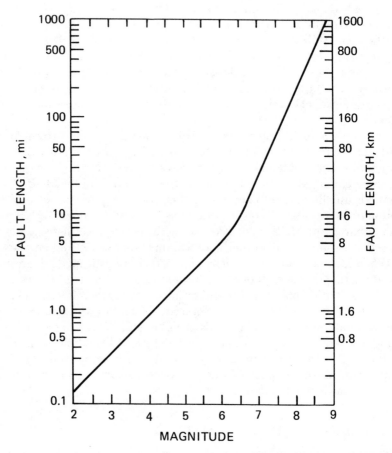

FIGURE 6-2   Idealized curve showing the approximate relation between the magnitude of the earthquake and the length of the fault rupture. For example, for the *M* 8.25 San Francisco earthquake of 1906, the graph gives approximately 250 miles for the length of fault slip, and this agrees with the observed length. For the *M* 6.5 San Fernando earthquake of 1971, the graph gives 10 miles, which is in good agreement with the length inferred after the earthquake. The graph is based on the assumption that for magnitudes equal or less than *M* 6 the slipped fault area is approximately circular in shape, although this is sometimes not true for real earthquakes. For large magnitudes in California the length of fault slip is large but the vertical dimension of fault slip is assumed not to exceed approximately 10 miles. Source: Housner and Jennings (1982).

cal value of earthquake magnitude must be selected from the data. To complicate matters, the faults that produce many of the earthquakes in the United States have not been identified; therefore, this method cannot be used in such cases.

Seismographs were invented as recently as the late nineteenth century, and the first magnitude scale was proposed by Richter in 1935. Therefore, magnitudes based on instrumental data can be assigned only to relatively recent earthquakes. However, the effects of earthquakes on people, structures, and land can be expressed in terms of earthquake intensity. (As noted in Chapter 3, in the United States the Modified Mercalli intensity scale is used for this purpose). For an individual earthquake the maximum value of intensity usually occurs near the epicenter and, thus, is called the epicentral intensity. Various empirical relations between epicentral intensity and the different kinds of magnitudes (e.g., local, body-wave, and surface-wave magnitudes) have been proposed. These relations show a dependence on geographical location, as well as on the strength of the earthquake.

For deterministic-statistical studies the distance from the earthquake to the dam site is taken as the minimum distance from the fault to the site. Because the actual earthquake may occur anywhere along the fault or in the source zone, this assumption can lead to overestimation of the ground motion at the dam site from any single event occurring on the fault. Over a sufficiently long period of time, motions associated with energy release on the nearest part of the fault can be expected to occur.

There are many proposed "attenuation relations," relations in the form of equations or curves that give an estimate of the strong ground motion if the magnitude, or epicentral intensity, and distance to the site are known. Examples for the western United States are shown in Figure 6-3. Because the fall-off of ground motion with distance varies geographically, different relations should be used for different regions. Thus, for any given region the data must be interpreted statistically, as is shown in Figure 6-4. For example, the attenuation is appreciably smaller east of the Rocky Mountains than to the west, resulting in larger felt and damage areas for eastern U.S. earthquakes. Most of the strong-motion data come from western U.S. earthquakes, for which empirical attenuation relations can be established. For the east, which is deficient in such data, various techniques that require additional assumptions must be used, which adds to the uncertainty of ground-motion estimates.

Finally, the variability of soil and poorly consolidated rock layers above competent bedrock can have an appreciable effect on ground-motion estimates. Sometimes a mathematical-physical model consisting of vertically propagating shear waves is used to estimate the local site effects. Although such a model is a gross simplification of actual conditions, it may provide useful insights into local site efforts. Alternatively, empirical correlations of

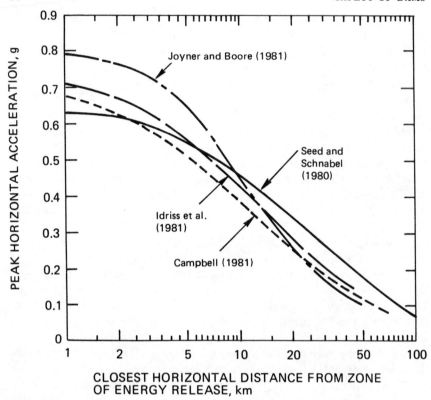

FIGURE 6-3 Peak ground acceleration curves for stiff soils ($M_s = 7.5$). Source: Seed and Idriss (1982).

ground motion for different soil conditions may be used, such as shown in Figure 6-5.

When all the uncertainties that appear in this method of estimation of peak ground motion are combined, the mean plus one standard deviation value may be almost twice the mean value.

SEISMOTECTONIC (SEMIPROBABILISTIC) METHOD

With few exceptions, earthquakes in the United States east of the Rocky Mountains cannot be associated with mapped faults. Although these earthquakes occur in the upper 25 kilometers of the earth's crust, the rupture planes do not extend to the free surface. As a consequence, fault rupture length cannot be determined from field evidence but rather must be inferred from characteristics of the earthquake wave spectrum near the source. This

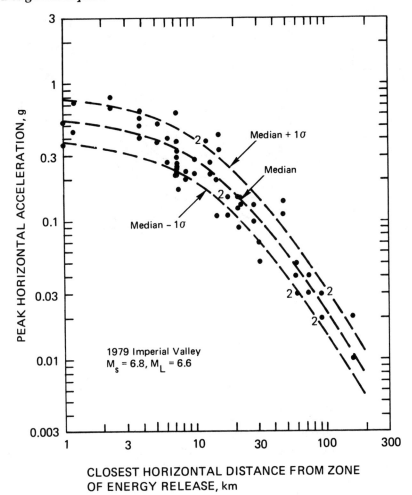

FIGURE 6-4 Regression analysis of the peak accelerations recorded during the October 15, 1979, Imperial Valley earthquake. Source: Seed and Idriss (1982).

adds further uncertainty into the relation between fault rupture length and earthquake magnitude, over and above that due to typical scatter of observational data.

When earthquakes cannot be associated with identifiable faults, the specification of distance from the earthquake to the dam site, as required for deterministic-statistical studies, takes on a significant amount of uncertainty. Accordingly, in the seismotectonic method the country or a portion of the country is divided into regions with similar geological and seismological

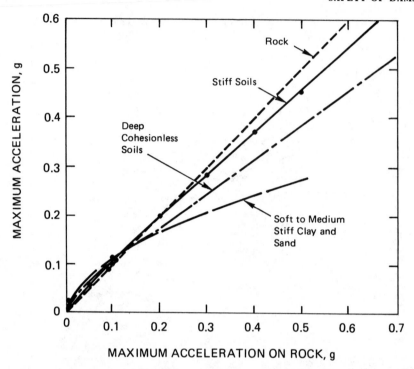

FIGURE 6-5 Approximate relationships between maximum accelerations on rock and other local site conditions. Source: Seed and Idriss (1982).

characteristics, and it is assumed that the spatial density of historical earthquakes is more or less uniform in each of these regions. Each such region is called a seismotectonic province or region.

An MCE must then be determined for each relevant seismotectonic province. Because usually the recurrence interval of earthquakes of that magnitude is much longer than the record of historic seismicity, the magnitude of the largest historical earthquake will be less than that of the MCE, which requires that probabilistic procedures be employed to arrive at an estimate of the magnitude of the MCE. Beyond this stage, the analysis proceeds as for the deterministic-statistical method, except that the distance from the earthquake source to the dam site is considered to be the smallest distance from any point in the seismotectonic province to the site, if the site lies outside the province. When the dam site lies within the seismotectonic province, the epicentral distance would become zero if the above criterion were applied, and the only attenuation of the strong ground motion would be due to the vertical travel path from the earthquake focus to the epicenter. This would result in unrealistically large estimates of the ground motion at the dam site, because the likelihood that the epicenter of the MCE would occur at the dam site is extremely small. Typically seismotectonic provinces in the eastern United States have dimensions of at least several hundred kilometers. Therefore, if the dam site lies within a seismotectonic region, it is normally considered to be sufficiently conservative to assume that the epicentral distance to the dam site is some small portion of the province dimension so that the probability of an epicenter being closer is sufficiently small.

PROBABILISTIC-RISK ANALYSIS

The deterministic-statistical and seismotectonic methods do not take into account the frequency of occurrence of earthquakes. Therefore, for areas where the MCE has a very long return period, deterministic-statistical estimates may lead to overly conservative estimates of the ground motions for structures that have lifetimes considerably less than the recurrence period of the MCE. In general, eastern U.S. earthquakes of any given magnitude have longer recurrence times than western earthquakes of the same magnitude, the difference being as great as 5-10 times. One of the principal purposes of applying risk analysis methods is to take account of the frequency of earthquake occurrence in hazard assessment. Also, the uncertainties in the various stages of calculation can, in a formal sense, be treated more readily in probabilistic methods. This latter advantage, however, may have little or no influence on the final selection of earthquake motions for which the safety of the dam is evaluated.

Similar to the deterministic-statistical and seismotectonic methods, the risk analysis method requires a knowledge of the location and extent of active faults or earthquake source zones, of the MCE associated with each, of the attenuation relations for peak ground acceleration and ground velocity, and of the site correction for soils and unconsolidated rock. In addition, the recurrence times for earthquakes of various magnitudes are required; oftentimes this relation is assumed to satisfy a simple mathematical form. Finally, possible variations in these parameters must be known. For example, it has been customary to assume that the earthquake occurrences are distributed randomly in time, resulting in a so-called Poisson distribution; however, physical reasoning suggests that this is not likely for the large-magnitude earthquakes that relieve most of the accumulated strain energy. Recently, other time distributions that take account of this phenomenon have been proposed and applied to a limited number of earthquake hazard calculations.

Rather than give a single estimate of the peak ground motion (e.g., acceleration) at a site, as is done with deterministic and seismotectonic methods, the risk analysis method gives a distribution of peak acceleration and velocity values at the site for various values of annual probability of exceedance. The smaller the probability value, the larger the values of peak ground acceleration and velocity. Alternatively, maps can be constructed that present the peak acceleration or velocity values that have a selected probability of occurrence in a selected number of years, e.g., a 10 percent probability of being exceeded in a 50-year time interval. The acceleration or velocity values on such maps can be contoured, as was done in those prepared by the Applied Technology Council (1978) and by Algermissen and Perkins (1976) and Algermissen et al. (1982) for the United States.

In the probabilistic method the uncertainties must be estimated for each step of the calculations and combined to give a mean ground-motion value and its standard deviation for each annual probability of exceedance. In general, the standard deviation increases significantly as the annual probability of exceedance decreases, particularly as the latter becomes less than about 0.001 or 10^{-3}. This is, in part, a result of the fact that the seismicity data base is known only for a few hundred years, at most, in the United States.

If the MCE motion at a dam site is to be determined solely by probabilistic-risk analysis methods in the future, there must be a decision as to an acceptable value of the annual probability of exceedance. That is, the acceptable amount of risk must be decided, with the realization that the standard deviation of peak ground acceleration and peak ground velocity values increases substantially as the annual probability of exceedance becomes less than 10^{-3} to 10^{-4} range.

OTHER EARTHQUAKE PARAMETERS

The foregoing discussion of deterministic-statistical, seismotectonic (semiprobabilistic), and probabilistic methods principally was concerned with the estimation of the maximum credible ground acceleration and velocity and their uncertainties. Because permanent displacement or failure of an embankment dam due to earthquake shaking may be the result of incremental slope failures or the consequence of liquefaction of the soil material comprising or supporting the dam, and because those effects are influenced by the number of cycles of strong ground shaking, the time duration of the maximum credible ground motion also must be estimated. Also, propagation of cracks in concrete dams is affected by the numbers of cycles of such shaking.

The time duration of strong ground shaking near the fault is largely a function of the length of fault rupture, with the duration time increasing as the rupture length increases. At large distances from the fault, attenuation results in a reduction of the amplitude of the strong ground motion. In addition, there is an effect, called dispersion, which causes the wave train to spread out in time as the distance from the earthquake source increases. Because of these complications, it is necessary to determine the duration of strong shaking by empirical means.

Different investigators use different definitions of duration time, so that a numerical value (usually given in seconds of time) is by itself meaningless unless the definition also is provided. To avoid this problem, a time history of ground acceleration for the maximum credible ground motion can be provided. Such a time history usually is an actual recording of ground motion, selected from a set of such recordings. Wherever possible, the selection of a time history is done on the basis of similarity of earthquake magnitude, distance to the site, and rock or soil conditions at the site.

One or more time histories of ground acceleration representative of the maximum credible ground motion can be provided for analysis. An accelerogram can be converted into a response spectrum, or a set of response spectra with different amounts of damping. Response spectra, after smoothing, can be used by the engineer for design proposes. Figure 6-6 shows an example of a design accelerogram (time history) and design spectra for Camanche Dam.

RESERVOIR-INDUCED EARTHQUAKES

The reservoir behind the 300-foot-high Koyna gravity dam in India started filling in 1962, and in 1963 a number of small-magnitude earthquakes occurred in the vicinity of the dam. As the depth of water in the reservoir increased in following years, the frequency of occurrence and the magnitudes of these local shocks increased. In 1967, six earthquakes of M 5.5-6.2 occurred, and on December 12, 1967, a damaging M 6.5 earthquake occurred within 3 kilometers of the dam. The strong shaking caused horizontal cracks at about two-thirds the height with slight traces of water leakage visible on the downstream face of the dam. The dam was located in a region of low historical seismicity (Zone O on the Indian seismic zoning map), so that the correlation of frequency of occurrence and magnitude with reservoir filling indicated a cause and effect relationship: filling of the reservoir presumably triggered stress failures (earthquakes) in a prestressed body of rock. Such presumed reservoir-induced earthquakes have been observed at the six dams listed in Table 6-1 and, in addition, smaller events have been observed at other dams. The Hsingfengkiang Dam, a concrete buttress structure, was cracked by the ground shaking in a 1962 earthquake. The dams listed in Table 6-1 were all high dams with deep reservoirs.

The occurrence of reservoir-induced earthquakes is a peculiar circumstance in which the building of the dam leads to the triggering of local

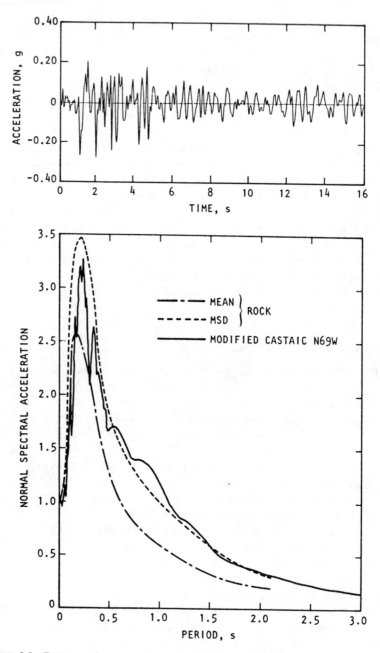

FIGURE 6-6 Design accelerogram and spectrum for Camanche Dam. Source: Seed and Idriss (1982).

TABLE 6-1   Dams at Which Apparent Reservoir-
Induced Earthquakes Have Been Observed

Dam	Location	Height (m)	Earthquake Magnitude
Koyna	India	103	6.5
Kremasta	Greece	165	6.3
Hsingfengkiang	China	105	6.1
Kariba	Rhodesia	128	5.8
Hoover	United States	221	5.0
Marathon	Greece	63	5.0

earthquakes of potentially damaging intensity. The first five dams listed in Table 6-1 were all located in regions of relatively low historical seismicity. The possibility of reservoir-induced earthquakes should therefore be given consideration when setting design criteria for new high dams, particularly in regions of low historical seismicity where the seismotectonic province method indicates a low intensity of shaking. At present, the committee is unable to assess confidently the likelihood of a reservoir-induced earthquake occurring at a proposed dam site.

7
Consideration of Risk in
Dam Safety Evaluations

INTRODUCTION

This report is concerned with risks arising from two types of events in the environment external to dams: extreme floods and earthquakes. Obviously dams and, consequently, the owners and others dependent on dams are subject to many sources of risks other than floods and earthquakes. A considerable number of these risks, including risk of dam failure from whatever cause, can lead to legal liabilities. The subject of legal liabilities and how they may be incurred is discussed in Chapter 8. This chapter compares the risks of dam failure with other man-made risks and includes discussions on the nature of risks from extreme floods and earthquakes, the attempts to cope with such risks, and how society has handled other types of risk with similarly potentially serious consequences.

RELATIVE IMPORTANCE OF RISK OF DAM FAILURES

There are some data that compare the impacts of dam failures to the impacts of other man-made and technological hazards in terms of a number of risk-related parameters. One study (U.S. Atomic Energy Commission, 1974) presents data in terms of annual probability of numbers of fatalities resulting from several man-made disasters, including dam failures. There it was concluded that deaths are considerably more likely to result from dam failures than from nuclear power plant disasters. Also in terms of 100 or fewer fatalities resulting from a single event, it was concluded that dam failures presented less threat than several other disasters, such as air crashes, fires, and explosions. However, it is noteworthy that of all the disasters on which data were presented for fatalities resulting from a single event, dam failures were shown to pose the greatest threat. Specifically, the study indicated that a dam failure causing 1,000 deaths might be expected on the average of once in less than 100 years. More recently, Christoph Hohenemser presented a discussion (Covello et al., 1983) on an approach to describing risk in terms of 12 dimensions of hazard for 93 types of technological hazards, including dam failures. Three of the dimensions are of particular interest: "population at risk" (i.e., people in the United States exposed to the hazard); "annual mortality" (average annual in the United States); and "maximum potentially killed" (maximum credible number of people that could be killed in a single event). Information presented indicates that the total population at risk from dam failures is in the same range (10 million to 100 million) as from several other hazards, such as fireworks accidents, skyscraper fires, train crashes, smoking, toxic effects from asbestos spray, and radiation from nuclear wastes. Further, information indicates that the maximum number of people who could be killed in a worst event is probably greater for dam failures than most any other kinds of hazards. Only a few hazards (principally those related to nuclear and war activities) are indicated to have potential for killing more people in a single event. Thus, many people—both individually and collectively—are *potentially* at risk from dam failures. This would seem to underscore the importance of good dam design, maintenance, and safety programs in a safety conscious society. Data on annual mortality may suggest that these objectives are generally being achieved.

According to Hohenemser, on the average, the number of deaths resulting from dam failures in the United States annually is in the range 10–100. This is in the same range as fatalities from dynamite blast accidents—where the population at risk is far fewer—and elevator falls—where the population at risk is greater. Many more deaths result from appliance accidents, commercial airline accidents, radiation from medical x-rays, train crashes, and about 30 other of those 93 causes tabulated. On the other hand, it is noteworthy that, according to Hohenemser, the annual mortality from bridge collapses (and a relatively few of the other hazards such as polychlorinated biphenyls, radiation from nuclear reactors, and a few others) is fewer than 10 in the United States, while the population at risk is greater than for dams. Others have noted that most catastrophic dam failures have been caused by site-specific factors that are not necessarily applicable at other sites. For such reasons it is difficult to compare historical records of fatalities resulting from dam failures with those brought about by other causes. While it is difficult to draw any conclusions relevant to the charge of this committee, it is clear (from these data and from our own intuition and experience) that dam failures represent a relatively low chance but great impact type risk to people and property. This low chance of failure can probably be attributed in general to good engineering design and construction of dams.

These data are designed to provide a rough comparison of hazards, probabilities of occurrence, and current outcomes across a number of areas. One cannot draw a direct conclusion regarding whether the risk management in one area is optimal or even satisfactory. The committee believes that risks should be managed by balancing the benefits of additional safety against the costs of achieving the lower risks. The above data contain no information on either the additional safety that would be possible in each area or the additional cost of enhancing safety. Thus, these data have no direct interpretation in terms of what would be optimal or even satisfactory safety goals for dams. However, they provide evidence of the outcomes of risk management in various areas and of what society seems willing to tolerate in terms of current hazards, probabilities of occurrence, and outcomes.

THE DESIGN PROCESS

The design of a structure (or machine) may be described simply and concisely as (1) finding the loads that the structure will bear and (2) proportioning the component elements to withstand those loads. This simple explanation may imply that the process is direct, the future loading can be determined, no judgment on the part of the designer is required, and the finished design involves no risks. This is not true. Few structures are so simple that the designer need not apply judgment. Also, the designer normally works with codes and standards that include allowances, based on past experience, for variations from the design model in such aspects as loadings, ultimate strengths, and workmanship and provide a factor of safety in all designs.

When it is attempted to design for extreme floods and earthquakes, the process becomes much more involved. At the present level of knowledge of extreme floods and earthquakes, the outstanding characteristic of such events is their indeterminacy. The only clues as to what may be expected in the future lie in man-made records and in physical evidence of past events, such as large earthquakes, extreme floods, and high rainfalls. But, whatever the future may bring, it will not exactly duplicate the past. From available evidence, estimates can be made of the probable maximum limits of future floods and earthquakes, but the size and timing of extreme floods and earthquakes cannot be certain. Hence, any such design involves an unknown, a risk factor.

As described more fully in Chapters 5 and 6, two basic approaches have evolved for providing estimates of extreme hypothetical flood and earthquake loadings. The deterministic approach is a procedure that seeks by analyses and reasonable combination of the causative processes to estimate the magnitude of a hypothetical flood or an earthquake at the dam site that has little or no chance of being exceeded. However, experience has shown that, as more data become available, estimates of such extreme events also change. The probabilistic approach seeks, by statistical study of past historical events, to estimate the return periods or annual probabilities of occurrences of extreme hypothetical flood or earthquake events of various magnitudes. Such estimates, also, have changed, sometimes radically, as more data have become available.

Both the deterministic and probabilistic approaches to establishing design requirements for floods and earthquakes have deficiencies. However, when considering resource allocations, the probabilistic method has one basic advantage: it furnishes estimates of frequency of occurrence of the design event. Of course, even if reliable estimates of probabilities of future flood or earthquake events at a dam site are established, there remains the problem of selecting the frequency of event appropriate for design.

COMPARISONS OF RISK MANAGEMENT STANDARDS

An attempt has been made to compare the current criteria for analysis of safety of dams against extreme floods and earthquakes with standards of other groups, particularly federal agencies, for management of other types of risks having similar potential social impacts. It was found that each such standard is so specific to the subject matter and practices of its particular fields that cross-discipline comparisons are difficult.

The federal government became active in risk management in a major way only recently. While some regulation of ship safety goes back more than a century, and the Food and Drug Administration (FDA) was created in the early part of this century, federal safety regulation is largely a product of the period since 1966. Just prior to that date, FDA was given a major increase in its responsibilities to actively regulate the safety of food and drugs. The year 1966 marked the creation of the National Highway Traffic Safety Administration. In short order there followed the creation of the Environmental Protection Agency (EPA), the Occupational Safety and Health Administration (OSHA), the Mine Health and Safety Administration (MHSA), and the Consumer Product Safety Commission (CPSC).

As discussed in *The Strategy of Social Regulation* (Lave, 1981), it seems that Congress, assuming that increasing safety would be easy and cheap, often has mandated that safety be achieved within a few years of creating a program and rarely thought about the cost of achieving safety. With a few exceptions, Congress does not specify the safety goal. One exception is the

Delaney clause of the Food, Drug, and Cosmetic Act, in which the goal is zero risk and is impossible to achieve.

The area of federal risk management has been characterized by controversy. Virtually every new regulation or agency decision is challenged in court, often with one party arguing the decision is too stringent and another party arguing that it is not sufficiently safe. This has led agencies to be intentionally vague about their safety goals; they have tried to avoid committing themselves or even being terribly specific about the goals for a specific decision. Thus, what follows is a review of staff practices, of particular standards, more than of official agency policy as set out in the *Federal Register*. This is particularly true for EPA and FDA.

There has been a recent review of agency attempts to comply with Executive Order 12291 requiring benefit-cost analysis of major (more than $100 million) agency decisions (Dower, 1983). Dower characterizes agency practice on assigning values to physical measures of benefits. While this is not directly part of agency safety goals, it is closely related. He reports that the Federal Aviation Administration (FAA) does explicit translation of premature deaths into dollars. Other agencies do some translation but do not use the dollar estimates in making official decisions.

The Nuclear Regulatory Commission went through a formal process to define safety goals for nuclear power plants. The agency formally adopted a goal that the risk of cancer in the most exposed population due to nuclear power would not be an increase in the cancer risk of more than 0.1 percent, or no more than one additional cancer in a background level of 1,000 cancers. This goal proved controversial in two senses. The first was that it is not clear how to translate the goal into individual engineering standards for nuclear reactors. The Nuclear Regulatory Commission hopes to slowly work through a process where this goal will be a direct guide to their regulatory staff. The second was that some consumerists claimed that this safety goal was insufficiently stringent. One of the commissioners pointed out this goal would sanction thousands of deaths due to nuclear power during this century, if many additional reactors are built and the exposed population is large.

Much of the agency risk statements are not goals so much as statements of what is a de minimus risk level. The Supreme Court vacated the OSHA benzene standard in 1980 on the grounds that OSHA had not found that occupational exposure to benzene constituted a "significant risk" at the prior standard. Reasoning that the law does not concern itself with trivia, the plurality of the court appeared to adopt a principal that would apply to all federal agencies: the agency must first find that the risk is "significant" before it can act. Accordingly, agencies have attempted to work toward a definition of what is a significant risk or what is a de minimus risk.

The FDA has had a difficult time with the absolute nature of the Delaney clause. To deal with contaminants in food colors, the FDA promulgated a rule that would allow carcinogenic contaminants if the resulting risk were small, perhaps one additional cancer in one million exposed people over their lifetimes. In general, this risk level of one in one million seems to be a sort of level to distinguish what is a negligible risk from one worth taking action on.

The EPA has adopted a similar approach. The Carcinogen Assessment Group has evolved rules within the group that specify a risk level of one in one million or one in 100 thousand as being the rule of thumb to distinguish a negligible risk.

The FAA specifies the failure rate for commercial aircraft components. Each component is to have a failure rate less than 10^{-13} per hour (9×10^{-10} per year). About 100 persons are killed in commercial airline crashes each year in the United States, although presumably, a small proportion of these are due to equipment failure, as distinct from human error.

When EPA enforces statutes for control of toxic substances and pesticides, the staff is instructed to balance the benefit of the product against the health risk. This leads to a much lower level of safety than is used for air or water pollutants under EPA statutes. Similarly, the FDA regulates drugs with the same sort of risk-benefit trade-off. If a drug is effective and there is no other effective drug that has less undesirable side effects, then the FDA will approve even drugs with extremely high risks, such as drugs for chemotherapy for cancer.

All of the agencies seem to require greater safety when many people could be killed at the same time. That is, they are more risk averse where many people are simultaneously at risk.

RETROFITTING TO MEET NEW STANDARDS

Many dam owners, including federal agencies, have found that dams built years ago fail by considerable margins to meet current agency standards for new dams. Many spillways at existing dams are deficient in light of such current standards. A much smaller but significant number of existing dams is suspected to present problems under earthquake loading standards currently used for design of new dams. No complete estimate is available for the cost of upgrading existing dams in the United States to meet current criteria for new dams, but it is evident that such costs could mount into the billions of dollars.

As noted elsewhere, as we continue to collect data on extreme rainfalls, floods, and earthquakes, we can expect our estimates of maximum events to be adjusted generally upward, resulting in even more dams that fail to meet the current criteria for new dams. Also, in general, the cost of retrofitting an existing dam to provide additional spillway capacity to pass a new design flood (as the result of a new probable maximum flood (PMF) estimate) can be expected to be higher than providing the same increase in capacity in a new dam. The same situation is usually found when considering upgrading an existing dam to meet current earthquake criteria. The question arises, then, whether safety standards for new dams should be applied to retrofitting existing dams. The problem is a very general one for risk management. New information can tell us that the risk of a technology is different from what we thought it was when we adopted certain criteria. Should this trigger corrective action for an existing structure? The answer ought to depend on the amount of risk and the cost of correction. The committee believes that risk management decisions should be based on a balancing of benefits and costs. Insofar as the costs of enhancing safety are much larger for existing dam than for one about to be built, this balancing would call for less safety in the existing dam. This is not to say that an unsafe dam would be tolerated, but that new dams would be designed to be "extremely" safe while existing dams were only retrofitted to be "very" safe.

How do other federal agencies deal with analogous problems? The answer is that all of them in fact distinguish between what is required of new installations and what is required in terms of retrofitting or remedial action. For various reasons, very few agencies have formal decision methods to apply for this purpose. In such decisions, government agencies are faced with problems of achieving balance between two social principles: equity and efficiency. Equity demands that all citizens be treated similarly. Efficiency demands that government not be unduly disruptive of legitimate actions of its citizens.

Peter Huber has examined the legal and regulatory aspects of this old-new risk situation in a perceptive manner (Huber, 1983). The following are extracts from his article in the *Virginia Law Review*:

Federal systems of risk regulation subtly but systematically distinguish the devils we know from the ominous unknown. An old risk-new risk double standard pervades regulatory statutes and decisions construing them. In a rough way the distinction between old and new risks makes good economic and political sense. Regulation of old risks presents problems and costs different from those encountered in regulation of new risks. In practice, however, the old-new division is usually ad hoc, inadequately developed, and inconsistently applied.

Risk-regulating statutes of all types share one common characteristic: they divide the regulatory universe between "old" and "new" sources of risk. What do "old" and "new" mean? For the present, a rough intuitive definition will suffice. Old risks are those to which society has been widely exposed before Congress or an agency finds federal regulation necessary. These risks are associated with products already on the market, with entrenched economic interests, or with an established technology. New risks loom on the horizon, threatening to undermine the perceived safety of the status quo. They include new sources of exposure to an old type of hazard, such as a new aircraft design, as well as risks associated with new technology such as nuclear power. Old risks are risks which society has already embraced or come to tolerate; new risks are those tied to unrealized opportunities.

If the difference between old and new risks is easy to explain, the cause of the systematic division of the two is not. The reasons underlying that division are a central focus of this chapter. Old risks derive from settled production and consumption choices and from established technology. Their regulation therefore often faces large economic and social obstacles and incurs transition costs. As the Food and Drug Administration (FDA) learned when it attempted to ban saccharin, old risks have identifiable and self-aware constituencies. New risks, on the other hand, may be regulated with less direct disruption of settled expectations. Their regulation incurs a different type of costs—lost opportunity costs. Lost opportunity costs are usually difficult to measure, and the bearers of these costs may be neither identifiable nor self-aware. As a result, the political costs of new-risk regulation may be comparatively low whether or not the economic costs of new-risk regulation are significant. Regulatory statutes thus systematically treat new risks more stringently than old ones.

Dividing the risk universe between old and new sources may seem reactionary, showing an irrational bias against technological change. Yet, the division grows from the usually correct assumption that transition costs are higher than lost opportunity costs. In addition, the division seems politically inevitable. Congress is simply unwilling to improve our risk environment without carefully attending to the impact on established expectations. On the other hand, Congress is quite willing to resist deterioration of that environment with disciplined firmness.

One agency, the Nuclear Regulatory Commission, did consider a formal criterion for addressing this problem when it was proposing its quantitative safety goals (U.S. NRC, 1981). In essence, it was suggested that all new nuclear reactors should be required to meet certain safety goals; however, when analysis of existing reactors showed the safety goals were not met, the required action would depend on the level of excess risk. While the proposal was not passed, it is described here as a unique example of one attempt to relate quantitatively relative levels of risk to required response. It was proposed that, if the risk exceeded the goals by a factor of 300 or more (e.g., goal of 10^{-5}, but indicated risk of 3×10^{-3}), immediate corrective action would have to be taken "within days"; where risk exceeded goals by a factor of 10–100, action must be taken "within months"; if by a factor of 3–10, action within years; and if by a factor of less than 3, action must be considered.

The Federal Aviation Administration comes closer to using a formal method than any other agency surveyed. If a risk is determined to mean a failure rate of 1 in 1 billion hours (1 in 114,155 years) or less, then it is considered extremely improbable or sufficiently remote not to take corrective action. For greater risks, action is determined by a benefit-cost analysis. Benefit-cost analyses could show that a new safety device makes sense on new aircraft but not on older aircraft because of the greater cost of retrofitting.

The Occupational Safety and Health Administration does not officially treat new plants and old plants differently. Obviously, it would be socially unacceptable for a federal regulatory agency to adopt policies that explicitly advocate allowing some workers, doing the same work for similar wages, to be regularly exposed to greater risks than other workers just because they worked in a plant that was more costly to make safe. In fact, however, when OSHA promulgates a standard (as they did for lead exposure), individual firms have managed to negotiate different phase-in schedules if they can show they are doing the best they can to come into full compliance. OSHA has also issued individual interim lead standards for specific smelters.

The Environmental Protection Agency also does not have a formal procedure for distinguishing between the new and the existing risks, although it is quite common for EPA to make such distinctions based on cost differentials. Thus, for example, emission standards differ for older and newer automobiles, and new source performance standards for power plants show a strong bias toward stiffer standards for new plants. Ethylene dibromide (EDB) was banned from further use in some products but different acceptable standards were applied to products containing EDB, varying according to their proximity to human consumption. Many other EPA examples could be cited.

By contrast, examples can be cited of situations where retrofitting is required if the danger is perceived as serious and immediate or if the cost of reducing the danger is low. Recall of automobiles to correct deficiencies related to safety, smoke detectors in residences, sprinkler systems for hotels, and correction of design deficiencies in commercial aircraft are some examples of such required retrofitting. The actions of the National Highway Transportation Safety Administration (NHSTA) illustrate that agency's approach to the problems in deciding when retrofitting should be required. NHTSA specifies safety standards to be applied to vehicles of a specified model year and thereafter. To date, NHTSA has never required manufacturers to recall and retrofit these safety features into existing autos. For example, seat belts were required in 1968 and subsequent models, but prior models need not be retrofitted. NHTSA must decide every time there appears to be a safety problem in a given model whether to require recall or to tolerate the problem in existing cars, because the expense of recall is too great, but must ensure that the problem is corrected in the subsequent production.

The Federal Energy Regulatory Commission (FERC) has addressed this problem as it relates to the higher estimates of probable maximum precipitation (PMP) contained in Hydrometeorologic Reports 51 and 52 (Schreiner and Riedel, 1978; Hansen et al., 1982) of the National Weather Service (see Appendix A). FERC does not require reevaluation of an existing spillway at a licensed project solely because of the higher PMP estimates if the following conditions have been met.

- A reasonable determination of PMP has been made previously.
- A probable maximum flood (PMF) has been properly determined.
- The project structures can withstand the loading or overtopping imposed by the PMF.

These examples suggest that different agencies handle the problem differently, that most of them do not have a general formal criterion for distinguishing risk acceptabilities, but that all of them do in fact recognize the need to be responsive to the greater costs of applying new safety standards to what exists than of applying these standards to what we do in the future.

A different approach to evaluating risk may sometimes be appropriate to decisions regarding an existing dam. Long-term experience with the type of dam involved or the functions it serves may indicate a good possibility that the dam will soon be abandoned and breached, or it may be replaced or rebuilt. Also, we may expect that technologies for evaluating dam safety and correcting deficiencies will continue to be developed. These considerations may sugggest that the primary determinate of need for upgrading the dam should be its probable safety over a relatively short time in the future, (say, over a 25- or 50-year period), rather than its safety over some indefinitely long period. Methods for determining probabilities of occurrence in definite time periods are discussed in Appendix D.

8
Risk and the Calculus of Legal Liability in Dam Failures

INTRODUCTION

In analyzing the potential legal liability for dam failure, a distinction must be made between federal and nonfederal structures. While the water escaping through a break in a dam does not concern itself with whether the dam is privately or publicly owned, the doctrine of sovereign immunity results in a different matrix of legal principles applying to federal flood control projects than for private facilities.

COMMON LAW AND SOVEREIGN IMMUNITY

It was clear under common law that the government could not be held liable for mistakes and errors that caused injury to others. By way of illustration, regardless of whether the government was negligent in designing, constructing, maintaining, or inspecting a facility, it would not be liable persuant to the doctrine of sovereign immunity. The doctrine, which originally meant "The King can do no wrong," precluded litigation against the sovereign, i.e., governmental bodies.

THE FEDERAL GOVERNMENT AND SOVEREIGN IMMUNITY

To a greater or lesser degree, all jurisdictions have abrogated sovereign immunity. On the federal level, Congress enacted the Federal Torts Claims Act in 1946, 28 U.S.C. 2671 et seq., to impose liability in torts cases.

84

However, there are several exceptions to the act. A major case illustrating the limits of liability is *Dalehite* v. *United States*, 346 U.S. 15 (1953), which involved a major disaster at Texas City, Texas. Ammonia nitrate, intended for use as fertilizer in rebuilding Europe pursuant to the Marshall Plan after World War II, exploded, resulting in extensive loss of life and property damage. The Supreme Court held the Federal Torts Claims Act did not include causes of action for strict liability or intentional misconduct. Thus, Congress had waived sovereign immunity only for acts of negligence. The case has since been reaffirmed.

Consequently, the only cause of action currently available to victims of a federal dam break is negligence. However, a separate statutory exception to the Act greatly limits relief in dam-break cases. This exemption was originally enacted in 1928 as part of the flood control project for the Mississippi River, and has since been reenacted in broad public works bills. Section 702c, (33 U.S.C. 702), the current version, provides

No liability of any kind shall attach to or rest upon the United States for any damages from or by flood waters at any place . . .

The germinal case in interpreting 702c is *National Manufacturing Co.* v. *United States*, 210 F.2d 263 (8th Cir. 1954), where the Kansas River in 1951 entered flood stage at Kansas City. Government officials allegedly released negligent information on the status of the river, which thereby precluded plaintiffs from transferring their movable property to safety.

The 8th Circuit concluded from the statutory language that Congress intended to safeguard the United States against liability "in the broadest and most emphatic language." Liability is precluded "at any place" and of "any kind" (Id. at 270). The basis of nonliability is public policy. Since the cost of flood control projects would be great, Congress did not want the inevitable flood damages to be part of the costs. Congress never intended to be an insurer of flood safety. In effect there has been a trade-off by Congress in that it has conditioned flood control appropriations upon freedom from liability:

Heretofore the great contribution of the United States to the struggle that has continued for generations and will long continue, to conquer floods, has been made on the basis of federal nonliability for flood damages. That has been the condition of the government's contribution (Id. at 275).

Thus, the purpose of 702c, and the judicial interpretation thereof, "was to place a limit on the amount of money Congress would spend in connection with flood control purposes." *Graci* v. *United States*, 456 F.2d 20, 25 (5th Cir. 1971). Since flood damage was sure to recur during the course of the extensive flood control construction program, Congress did not intend to burden its efforts by paying out damages. On the contrary, it refused to compensate

victims of every act of God disaster because of the enormous financial implications. *Guy F. Atkinson Co.* v. *Merritt, Chapman & Scott Corp.*, 126 F.
Supp. 406, 408–409 (N.D. Calif. 1954).

It should be noted that legislative history was of little help in construing
the statute since the provision was not introduced until shortly before the
final version of the Flood Control Act was passed. The only comment made
with respect to the provision was by a congressman on the House floor, "to
the effect that in engaging itself in flood control works, the government
should not by itself be open to suits for flood damage." *Graci* v. *United States*,
301 F. Supp. 947, 953 n.8 (E.D. La. 1969).

The Federal Torts Claims Act did not expressly include 702c among its
provisions. However, the courts have held that 702c survived the enactment
of the Federal Torts Claims Act because of the fundamental policy of 702c.
See, e.g., *National Manufacturing Co.* v. *United States*, supra, and *Clark* v.
United States, 218 F.2d 446 (9th Cir. 1954).

A long-settled public policy is not to be overridden by the general terms of
a statute that does not show with certainty a legislative interest to depart
from that policy. Another rule of statutory interpretation is also applicable
here. When the legislature becomes aware of the judicial gloss placed on a
statute, but fails to exact measures that would change this interpretation,
then the legislative "inaction" is viewed in effect as a ratification and approval of the prior judicial action. It should be noted that on several occa-
sions, such as the Teton Dam Compensation Act, Public Law 94–400, 90
Stat. 1211 (1976), Congress expressly had an opportunity to modify or repeal
702c, but failed to do so. Thus, 702c continues as a statement of national
policy. *Aetna Insurance Co.* v. *United States*, 628 F.2d 1201 (9th Cir. 1980)

Another reason exists for the broad interpretation given 702c by the
courts. Government projects cannot insure against flood losses. Since the
government cannot guarantee that its flood control works and that even
under the best design, engineering, construction, operation, and maintenance conditions can prevent all flood losses, there should be limits imposed
upon its liabilities.

Once the basic policy is established, the parameters become clear. The
protection extends to lands far away, as well as in close proximity to the dam.
Villarreal v. *United States*, 177 F. Supp. 879 (S.D. Tex. 1959) (farm was 65
miles from the river). It covers negligent construction or maintenance of
flood works. *Stover* v. *United States*, 332 F.2d 204 (9th Cir. 1964). It encom-
passes the peripheral aspects of a flood control project, such as the relocation
of railroad tracks. There is no protection against backwaters caused by
floodwaters held in a project, such as a bridge embankment, which is an
integral part of the project. *McClaskey* v. *United States*, 386 F.2d 807 (9th
Cir. 1967). It avails plaintiff nothing to label the flow as "rapid runoff of
surface waters" rather than "floodwaters," *Florida East Coast Ry. Co.* v.

United States, 519 F.2d 1184 (5th Cir. 1975), or to attribute the damage to seepage. *Morici Corp.* v. *United States*, 491 F. Supp. 466 (E.D. Calif. 1980), affirmed on other grounds, 681 F.2d 645 (9th Cir. 1982). Section 702c applies whether the dam is a U.S. Army Corps of Engineers project or a Bureau of Reclamation structure. See, e.g., *McClaskey* v. *United States, supra* (Corps of Engineers); *Sanborn* v. *United States*, 453 F. Supp. 651 (E.D. Calif. 1977) (Bureau of Reclamation). The key to 702c is simply whether the dam was authorized in whole, or in part, as a flood control project. See, e.g., *Graci* v. *United States*, 456 F.2d 20 (5th Cir. 1971). Cases have even held that no liability would attach on a flood control project even when government negligence has caused or aggravated the loss. This holding was based on a fact pattern involving an act of God. See, e.g., *Burleson* v. *United States*, 627 F2d 119 (8th Cir. 1980).

On the other hand, there have been cases in which liability has been imposed on the government despite 702c. These situations involve drainage facilities incidental to other government installations, or government conduct unrelated to congressional flood control acts or projects. One case imposed liability when negligence occurred in the construction of a navigation aid project. The government contended 702c afforded an absolute immunity from liability for floodwater damage regardless of whether or not government negligence was associated with a flood control project. The court rejected this contention, noting that 702c was enacted for the purpose of flood control projects. *Graci* v. *United States*, 456 F.2d 20 (5th Cir. 1971).

Liability was also found where, without warning, Air Force personnel at Ladd Air Force Base in Fairbanks, Alaska, dynamited an ice jam created by natural causes in the Chena River. *Peterson* v. *United States*, 367 F.2d 271 (9th Cir. 1966). Similarly, when negligent maintenance of a stream and culverts caused flooding, liability was found because an airfield construction project, and not a flood control project, was involved. *Valley Cattle Co.* v. *United States*, 258 F. Supp. 12 (D. Haw. 1966).

In conclusion, as one court has stated "It does not follow that the mere happening of a flood insulates the Government from all damage claims flowing from it." *Mc Claskey* v. *United States*, 386 F.2d 807, 808 n.1 (9th Cir. 1967).

THE RISKS OF NEGLIGENCE

The existence of risk is an integral component of any determination of legal liability. For example, negligence, which is the most commonly utilized cause of action both in tort litigation and dam failures, is generally defined in terms of the failure to exercise the standard of care of a reasonable person under similar circumstances. This standard in turn is based on the reasonable foreseeability of the risk. Determining whether this standard is met is generally based upon the risk that an accident will occur, the magnitude of harm should the risk materialize, and the availability of alternatives.

The classic formula was expressed by the distinguished jurist, Judge Learned Hand, in *United States* v. *Carroll Towing Company*, 159 F.2d 169 (2nd Cir. 1947) as follows.

Possibly it serves to bring this notion into relief to state it in algebraic terms. If the probability be called P; the injury L; and the burden B; liability depends upon whether B is less than L multiplied by P, i.e. whether B is less than PL.

It is important to emphasize that the ultimate question though is not foreseeability per se, but whether in light of that foreseeability, how a reasonable man would have acted taking into account the potential magnitude of harm, and the alternatives available. For example, if a specified flood were foreseeable, but highly improbable, should a dam engineer design the structures to handle that degree of flooding, or to a lesser standard? In this respect, if litigation ensued after a dam failure, both plaintiffs and defendants would introduce expert testimony on the standard of care to be expected under the circumstances. At that point the appropriate standard would be determined by the trier of fact, which is usually a jury.

If there is a recognized professional standard of care, then that standard will generally serve as the minimal legal duty. In this respect, if the Corps of Engineers' PMF spillway requirements are viewed as the appropriate standard for high-hazard dams, then that standard would most likely control the legal outcome.

Parenthetically, it should be noted that the question is not whether a similar event has occurred before, but the foreseeability of the risk that this particular mishap will occur. Even if a dam had not failed in the past under similar circumstances, liability may still exist if reasonable design, construction, operation, inspection, or maintenance procedures could have prevented the dam failure.

Because of the potential risks involved with a dam failure, the standard of care frequently imposed by courts is that one must use care commensurate with the undertaking, i.e., the duty of reasonable care is measured by the magnitude of the project. Obviously, the standard of care is a sliding one. While slight care might be required for a small stock-watering pond in an unpopulated, rural area, it would be grossly improper to use slight care in designing, constructing, or maintaining a large dam overlooking a major population area.

For example, Minnesota has held that since the standard of care is in proportion to the risk of injury, the owner must build a dam to meet such extraordinary floods as may be reasonably anticipated. *Willie* v. *Minnesota Power & Light Co.*, 250 N.W. 809 (Minn. 1933). See also *Herro* v. *Board of*

County Road Commissioners for County of Chippewa, 368 Mich. 263, 118
N.W. 2d 271 (1962). If the risk is high enough, the practical results approach
strict liability.
   As stated in the basic treatise in Tort law,

[I]f the risk is an appreciable one, and the possible consequences are serious, the
question is not one of mathematical probability alone. The odds may be a thousand
to one that no train will arrive at the very moment that an automobile is crossing a
railway track, but the risk of death is nevertheless sufficiently serious to require the
driver to look for the train and the train to signal its approach. . . . As the gravity of
the possible harm increases, the apparent likelihood of its reoccurrence need be
correspondingly less to generate a duty of precaution.

   Negligence can apply to the design, construction, operation, or maintenance of a dam. It can also consist in the failure to inspect a dam, or negligence in the actual inspection of the facility. Negligence can, thus, consist of
a failure to act or, if one has in fact acted, the failure to act in a reasonable
manner.
   It is also important to note that the higher the level of expertise, or degree
of training and expertise, the more one is held to a higher standard. For
example, if an emergency life-saving operation must be performed on the
side of the road, a general practitioner would not be held to the same level of
care as a general surgeon under those circumstances. Thus, an expert designing, building, or operating a dam will be held to the same degree of care as
other experts of the same background, training, education, and experience.
The expert will also have a duty to stay current in the field.

OPERATING DURING A FLOOD

   Another situation occurs when floodwaters pass through or over a dam,
flooding out downstream residents. The general rule in this country is that
the operator of a dam may permit floodwaters to pass over the dam in an
amount equal to the inflow, but will be liable if any excess amount is discharged. The basic premise behind the rule is that a downstream plaintiff
would have been damaged in any event by the flood, so he should not be
allowed to recover damages simply because of the fortuitous fact that a dam
was built, but did not have a sufficient capacity to capture the flood. It is,
therefore, assumed that defendant's acts did not in fact cause plaintiffs'
injuries since the damage would have occurred irrespective of the dam's
existence. Such a result can occur when a storm is of such intensity, as were
Hurricanes Connie and Diane in Connecticut in 1955, that plaintiff would
have been washed away regardless of a dam's existence. There is no legal

liability because there is no causation in fact. *Krupa* v. *Farmington River Power Co.*, 147 Conn. 153, 157 A.2d 914 (1959).

Consequently, it seems relatively clear that there is no duty on the part of a dam owner to operate the dam as a flood control mechanism for the benefit of lower riparian interests. Any cause of action must be based upon the negligent release of excessive water. The dam owner is essentially free to pass on the natural flow of the stream. See, e.g., *Baldwin Processing Co.* v. *Georgia Power Co.*, 112 Ga. App. 92, 143 S.E. 2d 761 (1965), *Crawford* v. *Cobbs & Mitchell Co.*, 253 P. 3 (Ore. 1927).

However, there is liability when a greater flow of water is released than is naturally flowing into the stream. This is especially true when "foreign" waters are being diverted into the reservoir. *Smith* v. *East Bay Municipal Utility District*, 265 p. 2d 610 (Cal. Ct. App. 1954).

There is some authority though, based upon the general duty of foreseeability of risk, that the operator of a dam has a duty to draw down a reservoir when heavy runoff is expected. See, e.g., *Kunz* v. *Utah Power & Light Co.*, 526 8 2d 599 (9th Cir. 1975). In this case the discharge did not exceed the natural flow of the stream. However, the operator had in the past skimmed the crest off of spring floods, thereby inducing a reliance expectation on the part of downstream farmers, who converted their crops from those that would survive flooding to those that would be damaged by flooding.

Similarly, in a case not involving a dam, *Salt River Valley Water Users Association* v. *Giglio*, 113, Ariz. 190, 549 P. 2d 162, 171 (1976), the court allowed recovery to homeowners who purchased homes in a floodplain. They successfully claimed that defendant's irrigation canal had inadequate spillways and, thus, caused flooding. There had been an unusual rainfall that approximated the 100-year flood. Liability was found even though the canal was not operated as a flood control device. Once the floodwaters entered the canal system, the association was under a duty to exercise reasonable care in disposing of that water.

As an added caveat, even if the operator is legally free to pass on the natural flow, there may be a duty to warn the downstream occupants of the high volumes of water that will be released. See, e.g., *Chrysler Corp.* v. *Dallas Power & Light Co.*, 552 S.W. 2d 742 (Tex. Ct. Civ. App 1975).

"ACTS OF GOD" AND PMF

A commonly asserted defense in dam failure cases is that the failure was caused by an "act of God," i.e., an eventuality outside human contemplation, such as a catastrophic storm. The act of God defense generally entails the following requirements: unforeseeability by reasonable human intelligence and the absence of a human agency causing the alleged damage. Thus, if a similar storm had occurred before, or could be anticipated using modern techniques, or if the storm or damage were otherwise reasonably foreseeable, even if not probable, act of God will not serve as a defense.

As explained in *Curtis* v. *Dewey*, 475 P.2d 808,810 (Idaho 1970), the "act of God" defense is based on the premise that

Negligence cannot be predicated upon a failure to anticipate that which was so extraordinary and utterly unprecedented as to have eluded the foresight of a reasonable man. If, therefore, a person builds a dam or embankment on or beside a waterway sufficient to withstand the maximum flow of water which might be expected, and the structure is destroyed by a flow which would not have been anticipated by a reasonably prudent man, then the resulting flood would be considered such an extraordinary flow of water as to amount to an "Act of God" and that person would not be negligent and not liable for damages caused by the flood.

A modern case, citing from an earlier 1916 opinion, laid out these factors in analyzing an act of God defense:

On passing upon what is or what is not an extraordinary flood or whether it should have been anticipated and provided against, the question to be decided is: "Considering the rains of the past, the topographical and climatic conditions of the region and the nature of the drainage basin as to the perviousness of the soil, the presence or absence of trees or herbage which would tend to increase or prevent the rapid running off of the water, would or should a reasonably prudent man have foreseen the danger and provided against it?"

Frank v. *County of Mercer*, 186 N.W. 2d 439, 443 (N. Dak. 1971), quoting from *Soules* v. *Northern Pac. Ry. Co.*, 157 N.W. 823, 824 (1946). While the defense has been successfully asserted in some cases, see, e.g., *Frank* v. *County of Mercer*, supra, it has received at best, a mixed reaction by the courts in dam failure cases.

A classic Colorado case illustrates the weakness of the act of God defense, and sheds some light on the current debate over the Corps of Engineers PMF requirements. In *Barr* v. *Game, Fish & Parks Commission*, 497 P. 2d 340 (Colo. Ct. App. 1972), design plans called for a spillway capacity of 33,000 cubic feet per second (cfs). The spillway constructed was for 4,500 cfs. The probable maximum flood was 100,000 cfs, although the previously known high flow of water was 27,500 cfs. The peak of the flood that occurred was 158,000 cfs with an estimated 75,000–100,000 cfs passing over the top of the dam. Defendants claimed act of God. The court rejected this defense, holding that the defendants were negligent in designing an inadequate spillway. Since the flow of water was reasonably foreseeable, there was no act of God.

The foreseeability of the risk (the probable maximum flood) was the key to liability. A similar result was reached in New Mexico, where the operator let sand and silt accumulate and failed to open a check gate. *Little* v. *Price*, 74 N. Mex. 626, 397 P.2d 15 (1964).

The act of God defense thus generally fails if the event should reasonably have been anticipated in light of past knowledge. While the past is prolog with respect to actually occurring events, foreseeability is based not only upon the historical past, but also what modern technology and science allows us to project into the future.

THE RISKS OF COMPLYING WITH MINIMAL GOVERNMENT OR PROFESSIONAL STANDARDS

It is also clear that compliance with a general accepted industry or professional standard of care, or with government regulations, establishes only the minimum standard of care. Courts may assess a higher standard of care, utilizing the "reasonable man" standard and the foreseeability of risk as the criteria. Judicial rejection of the governmental or professional standard does not occur as a routine matter, but it does occur often enough to transcend the unusual. It is fair to say that operators, who rely blindly upon a government or professional standard of care, are acting at great legal risk to themselves, when they *know or should know* that reasonable prudence requires higher care.

A good example of where compliance with a government standard was inadequate to preclude legal liability is *Gryc* v. *Dayton-Hudson Corp.*, 197 N.W. 2d 727 (Minn. 1980), where a 4-year-old girl received severe burns upon her upper body. She was wearing pajamas made of untreated cotton. The material did meet the federal standards of product flammability. The plaintiff established at trial that (1) the government standards were clearly inadequate at the time of the accident, (2) the apparel manufacturers were vigorously fighting any change in the government standards, (3) there were available commercially durable flame-retardant chemicals that would have significantly increased the safety of the product, and (4) the defendant was aware of those facts. Consequently, it was found that the defendant acted in reckless, wanton, and/or malicious disregard of the rights of others in marketing the fabric. The verdict of $750,000 compensatory damages and $1,000,000 punitive damages was therefore affirmed on appeal.

THE RISKS INHERENT IN DESIGN TRADE-OFFS

While in some sense there must always be a trade-off between absolute safety, performance (efficiency), and cost (economics), the practical reality is that, in the eyes of a jury mesmerized by a skillful attorney, trade-off will always seem callous when balanced against the lives lost or severely injured as a result of that decision. This exercise of discretion on the part of the designer or operator may well appear to constitute a "reckless disregard" for the rights of the victim, since the injury was foreseeable.

A good example is *Dawson* v. *Chrysler Corp.*, 630 F.2d 950 (3rd Cir. 1980). The plaintiff, a police officer, was rendered a quadriplegic when he lost control of his police car on a rain-slicked road and crashed into a telephone pole. The car struck the pole in a backward direction at a 45° angle on the left side of the vehicle. Point of impact was the left rear wheel well. The vehicle literally wrapped itself around the pole. The pole ripped through the body of the car and crushed the plaintiff between the seat and the "header" area of the roof. He claimed the vehicle was defective because it did not have a full, continuous steel frame extending through the door panels and a cross member running through the floor board between the posts located between the front and rear door of the vehicle. The plaintiff alleged that with such a design the car would have bounced off the pole with little injury to himself, who incidentally was not using his seat belt.

The plaintiff successfully recovered a verdict of $2,064,863.19 in spite of Chrysler's evidence that the vehicle met all federal requirements and that the plaintiff's design theory would create a greater risk of injury in most auto accidents. The Chrysler design in question absorbed the impact of most crashes (like an accordian) and decreased the rate of deceleration on the occupants of the vehicle. In addition, the plaintiff's design would add between 200 and 300 pounds to the weight of the vehicle and about $300 to the price of the vehicle. Yet the plaintiff won. The reason is obvious. It has to do with the risks of defendant going to trial with a severely injured victim for whom the jury understandably feels sympathy.

THE RISKS OF STRICT LIABILITY

The major alternative theory to negligence is strict liability. If such a theory is used, we realistically do not concern ourselves with the degree of care used by the defendant or how otherwise reasonable his conduct was. Strict liability essentially imposes liability as a risk of doing business.

Strict liability is derived from the old English case of *Rylands* v. *Fletcher*, L. R. 3 Eng. IR. App. Cas 330 (1868), where defendants constructed a reservoir on adjacent land in Lancashire with the owner's permission. Abandoned mine shafts underlaid the area, which is similar to the Scranton, Pennsylvania, region of the United States. Upon partial filling by defendants, the shafts gave way under pressure, causing water to flow into defendants' workings, and thence into plaintiffs', destroying them in the process.

The court ruled for plaintiffs, holding that when one brings onto his land, and collects and keeps there anything likely to do mischief if it escapes, and it is a nonnatural use of the land, he must keep it at his peril. If not, he is prima facie answerable for all the damages that are the natural consequences of its escape. As developed by the British courts, the rule is that the defendant is liable when he damages another by a thing or activity unduly dangerous and inappropriate to the place where it is maintained, in the light of the character of the place and its surroundings.

Rylands v. *Fletcher* initially met a lukewarm reception in the United States but has now become generally accepted. Critical in the early rejection of *Rylands* v. *Fletcher* was that the doctrine would have hindered an expanding civilization and industrialization. However, social values have changed over the past century. Today we have a fault system of liability, which is partially based upon the entrepreneurial risk of doing business. We also place more emphasis on victim compensation today and less on the economic needs of the defendant.

The concept of strict liability has been extended widely to activities considered abnormally dangerous or ultrahazardous. The basis of strict liability for ultrahazardous activities is the risk of harm and the potential magnitude of that harm should the risk be realized. In such a situation, liability does not depend upon such factors as intent, recklessness, knowledge, negligence, moral blameworthiness, or any other degree of culpability. Nor does it depend on the degree of care the defendant exercised. Rather, liability is based simply on the risks involved.

While strict liability for ultra hazardous activities has become widely accepted in the states, its application to dam failures has been more limited. There are not many relevant cases; most are older and several are based on policy considerations. For example, *Rylands* v. *Fletcher* was rejected by Texas in a famous case involving the escape of salt water from ponds constructed to handle the runoff from oil wells. It was technologically impossible to produce oil without drawing up salt water. Under the circumstances, the Texas Supreme Court did not want to hinder the oil industry. *Turner* v. *Big Lake Oil Co.*, 128 Tex. 155, 96 S.,W. 2d 221 (1936).

A slight majority of states reject strict liability in dam failures, including a relatively recent 1972 New Hampshire opinion. *Moulton* v. *Groveland Paper Co.*, 289 A.2d 68 (N.H. 1972). Several older California decisions also reject strict liability in dam failures. *Guy F. Atkinson Co.* v. *Merritt, Chapman & Scott Corp.*, 123 F. Supp. 720 (N.D. Calif. 1954); *Sutliff* v. *Sweetwater Water Co.*, 182 Calif. 34, 186 P. 766 (1920). In light of more recent California cases in other areas of the law, reliance on these older cases to limit liability is highly questionable.

More recent Massachusetts and Florida opinions accept the doctrine. See

Clark-Aiken Co. v. *Cromwell-Wright Co.*, 367 Mass. 70, 323 N.E. 2d 876 (1975) and *Cities Service Co.* v. *State of Florida*, 312 So. 2d 799, 801 Fla. App. (1975). The Florida case involved the breach of a phosphate settling pond, causing one billion gallons of phosphate slime to escape, "killing countless numbers of fish and inflicting other damages." The court adopted *Rylands* v. *Fletcher*, setting out policy grounds that are widely applicable today:

In early days it was important to encourage persons to use their land by whatever means were available for the purpose of commercial and industrial development. In a frontier society there was little likelihood that a dangerous use of land could cause damage to one's neighbor. Today our life has become more complex. Many areas are over crowded, and even the nonnegligent use of one's land can cause extensive damages to a neighbor's property. Though there are still many hazardous activities which are socially desirable, it now seems reasonable that they pay their own way. It is too much to ask an innocent neighbor to bear the burden thrust upon him as a consequent of an abnormal use of the land next door. The doctrine of *Rylands* v. *Fletcher* should be applied in Florida.

The Restatement of Torts essentially adopts *Rylands* v. *Fletcher* in imposing liability for ultrahazardous activities, which necessarily involve a risk of serious harm to others, which cannot be eliminated by the exercise of utmost care, and are not a matter of common usage. Factors to be considered include the high degree of risk, the potential gravity of harm should the risk materialize, the exercise of reasonable care, whether or not the activity is one of common usage, the appropriateness of the activity to the locality, and its value to the community. A reading of the cases indicates that the major factor is the nature and extent of the risk. This analysis, particularly the emphasis on risk, proved critical in the Massachusetts case of *Clark-Aiken Co.* v. *Cromwell-Wright Co.*, 367 Mass 70, 323 N.E. 2d 876 (1975), which adopted strict liability in dam failure cases.

Strict liability has also been imposed in situations where the defendant has constructed a dam, or part of a dam such as flash boards, expecting it to give way in a flood. In such a case, the potential risk of downstream flooding is so great that liability is imposed. It should be noted today that the operator under such circumstances could be considered "reckless" in his actions and, thus, potentially subject to punitive damages.

Occasionally a state will have a statute that imposes strict liability in dam failures. See Colo. Rev. Stat. Section 37–87–104, which provides, "The owner of a reservoir shall be liable for all damages arising from leakage or overflow of the waters therefrom or floods caused by the breaking of the embankments of such reservoir." New Hampshire has a statute that makes it unlawful to have a "dam in disrepair." N.H. Rev. Stat. Ann. Section 482.42.

Violation of the statute gives rise to civil liability. *Moulton* v. *Groveland Paper Co.*, supra. In this situation, the legal cause of action is technically negligence and not strict liability.

CONCLUSION

One added comment should be made here. It should be emphasized that tort law in general, whether the theory is negligence or strict liability, is moving in the direction of victim compensation. Consequently, as in *Dawson* v. *Chrysler Corp.* discussed above, most courts strain to invoke liability, particularly when personal injury or death is involved. The odds are substantial that regardless of the theory cited, the result will be a finding of liability in the case of a dam failure involving loss of life.

9
Proposed Hydrologic Criteria

BASIS FOR PROPOSALS

The selection of level of protection against extreme floods for a specific dam, like many design choices in engineering, is basically a problem in allocation of resources. However, as noted in Chapter 5, none of the currently available approaches to evaluating the safety of a dam against such floods provides a fully satisfactory method for this allocation. The deterministic approach, with its concentration on probable maximum events, does not directly consider problems of resource allocation. The probabilistic and risk-based approaches require estimates of probable frequencies of extreme flood events. Estimates of this type based on currently available data and techniques do not inspire high levels of confidence. Also, as noted in prior chapters, an analysis using the risk-based approaches may be nullified by changes in downstream development and other factors included in a risk-based analysis.

In recognition of these problems, the proposals set out in the following paragraphs seek to strike reasonable balances between what is theoretically desirable and what is practical based on current technologies.

SAFETY EVALUATION FLOOD

The committee recommends the adoption of the term safety evaluation flood (SEF) to designate the maximum flood for which the capability of the dam to withstand extreme floods without failure is to be determined. Such

97

usage will avoid the incongruity of using the term "spillway design flood" in connection with investigations of an existing dam when no design is contemplated. It also avoids the implication that only the spillway is involved in establishing the capability of a dam to withstand floods.

The selection of an appropriate SEF for a specific dam should consider that, as new information is collected, estimates of flood magnitudes and frequencies tend to change. As noted elsewhere, increases in estimates of probable maximum precipitation (PMP) have caused increases in estimates of probable maximum floods (PMF) in large sections of the country. Similar increases sometimes occur in estimates of magnitudes of floods of given frequencies as more data become available for frequency analyses.

PROPOSED CRITERIA FOR NEW HIGH-HAZARD DAMS

As noted in Chapter 7, there is a general tendency to impose a higher standard of safety on new developments that create new risks than is required of existing developments. There are reasons why higher standards might be imposed upon proposed dams as opposed to existing dams. First, for an existing dam the option of not building a dam in the first place is no longer available. A dam has been built, and all those living downstream of the dam are already exposed to some risk of dam break. Moreover, intentionally removing the dam to eliminate any possibility of breaching is usually not a tenable option because such removal would (1) increase the frequency of downstream flooding, (2) squander a valuable economic resource in which many may have invested, and (3) deprive many individuals of such benefits as recreation, irrigation, and water supply on which they have come to depend. Such constraints are not involved when a dam is proposed. Also, decisions in design for a new dam based solely on economic analysis without regard to who bears costs and risks could violate principles of equity. Such considerations indicate that the design of a new dam for a location upstream from an urban-type development, which would introduce potential for excessive damages and loss of life in the event of dam failure, should incorporate the maximum reasonable level of protection against failure during extreme floods, unless it can be shown that the failure of the dam during such floods would not increase the potential for loss of life and damages downstream.

Although legal considerations do not point to a specific basis for design of a new dam, court decisions have emphasized the need for dam designs to meet standards of reasonableness and prudence and to provide for reasonably foreseeable risks.

All things considered, the PMF provides the best criteria currently available to meet the standards of reasonableness mentioned above. Furthermore, although some may question justification for such conservative approach, the dam engineering profession has more confidence in the adequacy of the PMF criteria for major dams than in any other criteria that have been advanced. Hence, retention of the PMF criteria for design of spillways for new dams in high-hazard locations is generally recommended. However, we note that there may be instances when a smaller SEF is appropriate. In spillway design, the concern should be with the incremental damages associated with dam failure during an extraordinary flood. The failure of a small dam during a PMF event may have a negligible impact on the downstream hydrograph some distance from the dam. In such instances, SEFs smaller than the PMF would be appropriate. Likewise, it may be the case that downstream areas would already be flooded and evacuation activities completed before a PMF would overtop some reservoirs, resulting in dam breach. This is again an instance in which SEFs smaller than the PMF are appropriate based upon an incremental hazard analysis.

It is the committee's recommendation that for proposed high-hazard dams, the PMF should serve as the SEF unless risk analyses that examine the incremental impact of overtopping and dam failure during an extraordinary flood demonstrate that little or nothing is gained by such a high standard. In such instances, smaller SEFs should be adopted. A reasonable SEF would be the smallest value that ensures that a dam breach results in no significant increase in potentials for loss of life or major property damage.

CRITERIA FOR EXISTING HIGH-HAZARD DAMS

Prescribing an appropriate safety evaluation flood for an existing dam where failure could result in significant loss of life or property damage in downstream areas should involve a number of considerations. One approach, which some agencies have considered in past years, is to require that all existing dams in such high-hazard situations be capable of passing current PMF estimates. Such a requirement raises problems because of the following:

• In some instances, little additional safety would be provided by modifying a project to pass the most recent estimate of the PMF.

• For some existing dams it would be extraordinarily costly to modify the project to accommodate the full PMF.

• It is rather general practice in some other fields of endeavor not to require that a facility, designed to meet one safety standard or criteria, be retrofitted or modified to meet newly adopted criteria unless the existing facility is currently judged, given new evidence, to expose the public to unacceptably large and immediate risks.

These factors suggest that some criterion in addition to the PMF should be considered for the SEF for at least some existing dams. Whether an existing dam should be subjected to the same safety criteria as proposed new dams becomes a fundamental question in public policy, of balancing risks among various interests, as mentioned in Chapter 2.

If the PMF is ruled out, the following types of alternatives may be considered as bases for a safety evaluation for an existing dam:

• a flood having some selected estimated annual probability or average return period;

• a project-specific evaluation based on trial analyses of effects in downstream areas of potential dam failures during floods of various magnitudes (this risk analysis approach is discussed further below); and

• a flood that is some arbitrary fraction of a PMF (or perhaps derived from a fraction of the PMP) or of a flood having some estimated probability.

As noted in Chapters 4 and 5, although arbitrary criteria of the last listed type are in use by some agencies, there are significant disadvantages to use of criteria based upon fractions of a PMF or fractions of a flood of certain estimated probability or combinations of such floods.

Probability-based criteria, as noted earlier, offer advantages in any type of risk-cost analyses or comparisons of various types of risks. However, our limited abilities to reliably assign probabilities to rare floods have restricted the usefulness of the probability approach. As noted in Chapter 2, even if it were possible to accurately predict the probabilities of all sizes of floods at a dam site, there would still be the problem of selecting an appropriate basis for testing dam safety. There is little in the way of general guidance or precept for directly choosing an SEF for a high-hazard dam based on estimated frequencies. California specifies the 1,000-year return period flood for testing small dams in remote farm areas. The Institution of Civil Engineers in the United Kingdom recommends use of the 10,000-year return period flood or one-half PMF (whichever is larger) for design of a dam where failure of the dam will endanger a community but rare overtopping is tolerable.

From the above it is apparent that there is not one universally satisfactory approach to establishing spillway capacity criteria for existing high-hazard dams. There are some dams where the additional damage and loss of life caused by a dam failure due to overtopping may justify protection for the full probable maximum flood. Other situations may indicate that protection against the PMF is desirable but compromise on such items as freeboard allowance could be tolerated. At present, the best-attainable flood frequency estimates for streams in the United States cannot be used directly to determine spillway capacity requirements with confidence that future experience will not greatly change the estimated probabilities. However, as discussed below, even crude extrapolations of flood frequency curves may give satisfactory bases for comparisons of alternative modification plans by use of risk analysis procedures. Thus, the use of risk-based analysis should be considered in safety evaluation of any existing high-hazard dam for which the PMF is not required.

As suggested in Chapter 7, for existing dams the primary need in an evaluation may be to estimate the probable safety of a dam over a relatively short time in the future. Appendix D outlines procedures for estimating probabilities that events will occur within definite time periods once estimates of the annual probabilities or average return periods are established.

As indicated by the above, the committee considers that there is no single, universally correct approach to evaluating the safety of all existing high-hazard dams against extreme floods. The characteristics of each such dam, its drainage basin, the purposes served by the project, the area that would be affected by dam failure and the development in that area should be considered in arriving at an appropriate SEF for the project. As a preliminary guide to such consideration the following sequence of activities is suggested for an existing dam that is expected to remain in place for an indefinitely long time:

• Develop the estimated PMF for the site. (It is considered that an estimate of the probable maximum flood potential of the watershed should be available for every dam classed as high hazard.)

• If it is reasonably probable that the dam would fail if overtopped and the incremental impact (marginal damages and potential loss of life) clearly would be of such magnitude that potential for overtopping must be eliminated insofar as reasonably possible, adopt the PMF as the SEF and proceed to develop any needed remedial measures to assure that the SEF may be safely passed with normal allowances for freeboard, etc. (In some situations encroachment on the normal freeboard allowance by the SEF may be considered as acceptable.)

• If the dam would be overtopped and probably fail during a PMF but it is not clear that remedial work to permit safe passage of the PMF is justified, determine the magnitude of the following floods:

(1) The maximum flood that can be passed by existing project works with little or no danger of dam failure.

(2) The minimum flood for which failure of the dam would cause no significant increase in downstream damages under present and foreseeable future conditions.

If the flood determined by (1) is larger than that determined by (2), consider the consequences of dam failure and loss of project services at a probable frequency indicated by flood (1). If it were judged that such risks can be tolerated, no remedial work to provide further safety against extreme floods would be indicated.

• If the flood as determined in (2) above is larger than that determined by (1) or if it is considered that the consequences of dam failures caused by a flood such as determined by (1) are unacceptable, proceed with a risk-based analysis such as discussed in Chapter 5 to develop further bases for decisions on remedial work.

CRITERIA FOR INTERMEDIATE- AND LOW-HAZARD DAMS

Safety evaluations for intermediate- and low-hazard dams are primarily concerned with the economic effects of their potential failures. However, a continuing problem in such evaluations is the actual or potential development of the area downstream from the dam after the dam is constructed and the consequent change in the hazard ratings for the project. For this reason any agency having responsibility for protecting the public interest in dams should require periodic critical review of the hazard ratings for dams previously rated as intermediate and low hazard.

It is noted that dams having intermediate- and low-hazard ratings do not occupy a prominent position in the programs of the agencies requesting this study. Hence, no specific recommendations for safety criteria for these classes of dams are presented. However, as noted in Chapter 5, some standardization among the agencies concerned with such dams in regard to classification and safety criteria would be desirable and is encouraged.

RISK-BASED ANALYSES

Risk-based analysis when used to determine spillway capacity requirements provides the opportunity to weigh objectively the relative merits of alternative modifications embodying either variation in scope or variation in design concept. The difficulties associated with this approach relate to (1) uncertainty associated with the probability assigned to floods in excess of the 100-year average return period and (2) the inability to place monetary values on such intangible considerations as the loss of life. The difficulties associated with the uncertainties in assigning probabilities for remote flood events can be partly overcome by performing sensitivity studies as part of the risk-based analysis. Also, the potential for loss of life can be quantified in a risk-based analysis but this loss of life aspect should not be combined with economic considerations.

Despite the foregoing difficulties, the committee endorses the basic concept of the risk-based method for some purposes. The method appears especially appropriate for examining alternative procedures for upgrading spillway capacities of existing significant and high-hazard dams.

At this time the committee would caution against strict adoption of a "benefit minus cost" rule or a "benefit over cost" rule. The reasons are severalfold, but relate to the uncertainty in assigning probabilities and the obvious inability to quantify many of the broad social issues encountered, which can range from the possible loss of life to environmental concerns.

Instead, the committee recommends initial ranking of the alternatives by the average annual value of the tangible costs. The latter should be the sum of incremental downstream or upstream damage caused by project operation and/or failure, cost of the proposed modifications, damage sustained by the dam and appurtenant structures, and cost of interrupted dam services. This initial analysis should be accompanied by a descriptive appraisal of the other, i.e., nonquantifiable (intangible), considerations that need to be brought to the attention of the decision makers and the public at large.

The foregoing procedures inherently involve the following considerations. First, the absolute value of the economic calculations could differ appreciably as procedures for determining the probabilities for flow between the 100-year and the PMF levels are varied. This explains the committee's reluctance to utilize specific benefit-cost rules. Second, although the relative cost positions in a ranking of alternative modification schemes may be affected by the frequencies assigned to extremely rare events, sensitivity studies should bring out this dependence and provide basis for judgments among the schemes. For the present, the committee recommends development of the frequency curves for average return periods in excess of 100 years in accordance with procedures described in Appendix E.

HAZARD CLASSIFICATIONS FOR DAMS

Dams are often categorized as high, medium, or low hazard, depending upon the potentials for loss of life and property damage existing downstream. Such hazard classifications are extremely useful for identifying dams whose failure due to earthquakes, to floods, or to structural, piping, or foundation problems could cause major losses. However, in contrast to dam failure due to other causes, dam failure due to extreme floods only occurs during flood events. Thus, the important dimension in terms of spillway design is the *incremental* loss of life or property damage that would result from dam failure when the river is already in flood. In many situations, low-lying buildings, camp sites and picnic areas, and residences in the flood-plain, will have been flooded and evacuated before peak flow rates are reached and a large dam might fill and its emergency spillway fail. Thus, one should not confuse the loss of life and property damages likely to occur from a "sunny day" dam failure, perhaps due to an earthquake, with the incremental aspect of the loss of life and property damage due to dam failure caused by an inadequate spillway during a major flood.

The considerable differences in hazard classifications in use by various agencies are discussed in Chapter 5. However, as such classifications are not of major importance in the programs of the agencies requesting this study, no specific classification table for categorizing dam hazards is proposed. It is suggested, however, that existing standards should be improved.

IMPACT OF PROPOSALS

Federal Implications

Part of the committee's charge was to comment on how its suggested methods and criteria would impact federal costs. However, not enough information is available to the committee to allow other than a few comments based on value judgments.

The suggested criteria for design of new dams appear to be generally in line with present application. Accordingly, no significant fiscal impact should result from these recommendations. However, if as advocated by the committee, it is no longer required that all existing dams meet current criteria for design of new dams, significant savings in costs of rehabitating existing dams should be achieved.

Where risk-based analyses are utilized, it is quite clear that the costs of analysis and design could increase significantly. However, it is believed that application of this method will lead to significant reduction in modification costs in certain cases. Therefore, the *overall* federal cost should be reduced.

Nonfederal Implications

This report has been prepared in response to a specific request submitted by two federal agencies and with primary focus on federal projects. As the economic analyses for most federal dam projects attempt to evaluate all project-related costs, regardless of who bears such costs, the use of risk-based cost analyses is pertinent to selection of levels of risk that can be tolerated at such projects. Past congressional actions in compensating damages resulting from failures of federal dams reinforce this view. Hence, the committee advocates use of risk-based analyses for federal dams in the face of acknowledged need for added research and development on this method of analysis. Similarly, the nonfederal owner is confronted by the obvious trend toward full liability for damages, as demonstrated by recent court decisions exhibiting strict liability concepts. However, there are major differences between the federal government and most nonfederal dam owners in their capabilities to sustain a major loss resulting from a dam failure. Accordingly, at this time, consideration of the public interest may cause these federal and state agencies that regulate dams in the interest of public safety to delay in adopting the risk-based analysis methods, although some state agencies already permit such practices. In the meantime, nonfederal owners, regulators, and designers will need to keep abreast of potential research findings relating to risk-based methods. Until the risk analysis approach can be extended to the nonfederal field, the potential overall savings to society that it appears to offer will not be realized.

10
Proposed Earthquake Criteria

BASIS FOR ANALYSIS

As indicated in Chapter 6, at the current level of knowledge of earthquake occurrences in the United States, any estimate of annual probability or probable average return period for a major earthquake at a specific site is subject to considerable uncertainty. However, there are indications that, in most areas, earthquakes of the order of magnitude of the maximum credible earthquakes adopted for safety evaluations may have average return periods of, say, 500 to 1,000 years, or more, depending on the importance of the project. Earthquake experience along some major fault systems suggests shorter average return periods. It seems clear that the earthquakes adopted for safety evaluations do not represent as rare phenomena as do probable maximum precipitation estimates, which are generally considered to represent rainfalls having average return periods of from 10,000 years to as long as 1,000,000 years. Of course, in some locations there is the possibility that there will never be further movement along a fault that has been judged potentially active.

Other considerations apply to any comparison of criteria for safety of dams against floods and earthquakes. All dams are exposed to threat of extreme floods and require provision for such events, even though it may be extremely unlikely that a specific dam will ever experience a flood of the magnitude assumed for its safety evaluation. By contrast, only for dams located in active seismic zones are potential earthquake motions considered in safety evaluations and design, and, as indicated above, an individual dam

106

is more likely to experience the safety evaluation earthquake than it is to experience the safety evaluation flood. Also, failure of a dam as a result of an extreme flood would only add to the damages caused by the flood without the failure. But dam failure during an earthquake is apt to be a "sunny day" event for which all damages are attributed to the failure.

In view of the above, the committee considers there is little basis for using less conservative criteria for any situation where failure of a dam under earthquake loadings could involve large damages and loss of life. Such comparisons of estimated frequencies as the above can raise questions regarding the major differences in probabilities of occurrence of design floods and design earthquakes. However, since both types of estimates represent probable maximum events, these differences in probabilities do not represent inconsistent approaches.

DESIGN CRITERIA

The committee suggests that the federal agencies adopt the terminology "safety evaluation earthquake" (SEE). The SEE is defined as the earthquake, expressed in terms of magnitude and closest distance from the dam site, or in terms of the characteristics of the time history of the free-field ground motions, for which the safety of the dam and critical structures associated with the dam should be evaluated. For high-hazard and functionally essential dams the SEE will be the same as the maximum credible earthquake (MCE) to which the dam will be exposed. For lower-hazard dams the SEE may be less severe than the MCE.

Since the flooding caused by an earthquake-produced dam failure can rapidly affect a large area and population, and the public expects a high degree of protection from dams, the earthquake safety evaluation criteria should be stringent and conservative. The committee concludes that the safety evaluation earthquake criteria for high-hazard dams should be as follows.

• The dynamic response of the dam to a safety evaluation earthquake ground motion, attenuated from the source to the dam, should be such that it does not result in loss of the reservoir.

• Some damage to the dam and its appurtenant structures not critical to the stability of the dam may be allowable.

• For earthquake safety evaluations the assumed level of water in the reservoir usually should be the normal full pool.

The committee concludes that a safety evaluation earthquake ground motion developed by the deterministic-statistical method applied to known causative faults, which is based, in part, on subjective evaluation, provides an appropriate level of conservatism for dam safety analyses. The committee suggests that federal agencies adopt this approach whenever possible. However, where earthquake sources are not well defined, the seismotectonic province (semiprobabilistic) approach should be adopted. Other alternatives are not considered to be sufficiently well developed to be practicable at the present time and may not be so in the near future.

The committee endorses the concept of an operational basis earthquake (OBE) ground motion, which is defined as the intensity of ground shaking that has a significant probability of occurring in a period of about 200 years. The dam should withstand this loading with no significant damage. The selection of an OBE is a matter of economics and public policy, and these might indicate the use of a period greater than 200 years.

DESIGN APPROACH

The committee endorses, in principle, the following approach to dam safety analysis and seismic design of dams currently being used by the federal agencies.

Concrete Dams

• Adopt "defensive measures" in design and construction that will ensure foundation and abutment integrity, good geometrical configuration, and effective quality control.

• For gravity and buttress dams in areas of low seismicity, pseudostatic analysis methods may be used to check safety against sliding and overturning; however, for high dams it may be appropriate to use more accurate methods.

• For high-hazard dams in areas of significant seismicity, dynamic analysis methods should be used for the analysis of structural response and induced stresses. For preliminary safety assessments, simplified methods may be used.

Embankment Dams

• In design and construction, adopt such "defensive measures" as ample freeboard, wide transition zones, adequate compaction of materials in foundations and embankment, and a high level of quality control.

• For reasonably well-built dams on stable soil or rock foundations, the pseudostatic method of stability analysis may be used if estimated peak ground accelerations are less than $0.2\,g$.

• In areas where peak ground accelerations may exceed about 0.2 g, for dams constructed of or on soils that do not lose strength as a result of earthquake shaking, a deformation should be estimated using dynamic deformation analysis techniques.

• For dams involving embankment or foundation soils that may lose a significant fraction of their strengths under the effects of earthquake shaking, a dynamic analysis for liquefaction potential, or strength reduction potential, should be performed.

COMBINED EARTHQUAKE AND FLOOD CRITERIA

The committee does not consider it necessary to design dams to withstand the simultaneous occurrence of the safety evaluation earthquake and the safety evaluation flood. However, if an existing spillway is subject to frequent use, and an analysis with an OBE indicates that the spillway may be so damaged as to be unusable, the safety of the dam should be considered unsatisfactory. For especially high-hazard dams, more stringent combinations of earthquake and flood criteria may be justified.

11
Continuing Development of Hydrologic and Earthquake Engineering Technologies

OVERVIEW

As noted in Chapter 3, the methods in use for estimating the hazards to dams that may result from extreme floods and earthquakes are based on relatively new and still developing branches of technology. Because so many agencies and individual scientists and engineers are involved in problems related to extreme floods and earthquakes, further development is confidently expected in our understanding of the natural phenomena involved and in the ability to analyze the effects of such events on dams and to design for such effects. Occasional failures can be expected that will focus public attention on dam safety problems and the need for further improvements. Such developments in many areas of science and engineering improve the ability to design and construct dams that can meet the tests of extreme floods and earthquakes. Thus, the safety criteria developed today may not be appropriate in the future. They should be reviewed and updated periodically by a committee consisting of designers, hydrologists, meteorologists, seismologists, engineers, economists, and representatives of the general public. In the remainder of this chapter, some areas of research and development are pointed out that are most closely related to engineering for the effects of extreme floods and earthquakes.

With present knowledge of meteorological, hydrological, and seismological phenomena, one can only regard the occurrence of a large earthquake or an extreme flood at a given site as a random event. It can never be certain just how big an earthquake or flood a dam will experience during its useful life.

Thus, a dam designer or an official responsible for balancing the interests of all segments of the public usually cannot prove conclusively that the provisions in a dam design for extreme floods and earthquakes are adequate and have resulted in efficient use of resources. As discussed in earlier chapters, this lack of certainty in project needs has been met in two ways: (1) by trying to establish probabilities for these extreme natural events and then selecting a probability considered adequate as a design basis or (2) by attempting to define the maximum flood or earthquake at the site that conforms to our present knowledge. As has been noted, each approach offers certain advantages and disadvantages, but neither may provide the basis for complete confidence in a dam design by all concerned. The broad aim of research and development efforts related to dam safety should be the improvement of this confidence factor.

At present the methods available for acceptably thorough analyses of a dam's safety from extreme floods and earthquakes require substantial commitments of financial and technical resources. To many owners of dams and to many agencies responsible for safety of many dams, these costs in technical manpower and in dollars are so high that they have delayed or prevented action to protect the public from the dangers of unsafe dams. Thus, appropriate research and development efforts should focus on development of simplified evaluation methods that could be applied with confidence with defined limits.

HYDROMETEOROLOGICAL RESEARCH

For the past several decades, most areas of meteorology, hydrology, and hydrologic engineering related to dam design have received attention in research programs. Thus, the field has not been neglected, but it is felt that attention to the research needs in these areas should continue.

Research in methods of assessing probabilities of extreme floods has not received much attention. The ability to assign probabilities with confidence to such rare events would greatly enhance the capability to make rational decisions on allocation of resources in dam design and construction. Past experience has shown the severe limitations of methods of estimating probabilities of future extreme floods based on stream discharge records at specific sites. However, it appears that the considerable collection of data on major storm rainfalls developed by the National Weather Service (in cooperation with other agencies) for studies and estimates of probable maximum precipitation could provide a basis for generalized estimates of probabilities of extreme rainfalls, which could then be the bases for estimates of probabilities of extreme floods in specific drainage basins. Research in this area, possibly by the Hydrometeorological Branch of the National Weather Service, should be considered. It appears that use of paleohydrological methods could offer methods of assessing magnitude of prehistoric great floods in some geographic areas. Further development of this approach should be considered.

Finally, the committee suggests that the Corps of Engineers and the Bureau of Reclamation and other agencies in the water resources research field pursue a program of research and development designed to upgrade our ability to implement risk-based analyses. This research can and should range from the appraisal of issues relating to the probability distribution of very rare events to the analysis of procedures for implementing reliable flood-warning and flood-fighting techniques.

EARTHQUAKE RESEARCH

Research relevant to earthquake effects on dams has made rapid strides in the past 20 years. Knowledge of where and why earthquakes occur has been much expanded, and knowledge of earthquake ground shaking and the consequent vibrations of structures has been markedly advanced. Continuation of this ongoing research will contribute to improved hazard assessment and to improved methods of seismic design of dams. However, earthquake research, in general, has not been particularly directed at seismic problems of dams, though much of it does have a bearing on these problems, for example, compilation of earthquake statistics, study of source mechanisms, recording of strong earthquake ground shaking, and development of powerful methods of computing the dynamic response of structures. It is felt that much more could be gained through research concentrating on the seismic problems of dams. Such studies include research on the nature of ground shaking at dam sites as affected by topographic and geologic features peculiar to dams and reservoirs; improved methods of identifying active faults and estimating the frequency with which they generate earthquakes; the dynamic response of dam-foundation-abutment-reservoir systems; generation of surface waves in the reservoir and behavior of pressure waves and their interaction with the dam and reservoir bottom; dynamic performance of concrete dams and their ability to withstand high tension stresses; capability of three-dimensional, dynamic analysis of embankment dams to predict permanent seismic deformations; earthquake sensors on selected dams to record the actual behavior during earthquakes; and additional strong motion instrumentation networks to further develop attenuation relationships for the various seismotectonic regimes, especially in the central and eastern United States.

APPENDIXES

APPENDIX A

Design Criteria in Use for Dams Relative to Hazards of Extreme Floods

CONTENTS

PART 3—OTHER GOVERNMENTAL AGENCIES

PART 4—TECHNICAL SOCIETIES

PART 5—FIRMS IN UNITED STATES

PART 6—OTHER ENTITIES IN UNITED STATES

PART 7—FOREIGN COUNTRIES

PART 1—FEDERAL AGENCIES

Ad Hoc Interagency Committee on Dam Safety of the Federal Coordinating Council for Science, Engineering, and Technology

This group, a forerunner of the present ICODS, issued "Federal Guidelines for Dam Safety," dated June 25, 1979. The following is extracted from those guidelines:

The selection of the design flood should be based on an evaluation of the relative risks and consequences of flooding, under both present and future conditions. Higher risks may have to be accepted for some existing structures because of irreconcilable conditions.

When flooding could cause significant hazards to life or major property damage, the flood selected for design should have virtually no chance of being exceeded. If lesser hazards are involved, a smaller flood may be selected for design. However, all dams should be designed to withstand a relatively large flood without failure even when there is apparently no downstream hazard involved under present conditions of development.

Bureau of Reclamation, U.S. Department of the Interior
(From letter dated June 6, 1984)

The following is extracted from a description of the Bureau of Reclamation's practices relating to floods and earthquakes:

The PMF (Probable Maximum Flood) is a hypothetical flood for a selected location on a given stream whose magnitude is such that there is virtually no chance of its being exceeded. It is estimated by combining the most critical meteorologic and hydrologic conditions considered reasonably possible for the particular location under consideration. The term PMF has been adopted by the Bureau which brings us in line with terminology used by all other Federal agencies. Many past Bureau publications use MPF (Maximum Probable Flood) which has the same definition and usage as the PMF.

Bureau of Reclamation procedures estimate the PMF by evaluating the runoff from the most critical of the following situations:

1. A probable maximum storm in conjunction with severe, but not uncommon, antecedent conditions.

2. A probable maximum storm for the season of heavy snowmelt, in conjunction with a major snowmelt flood somewhat smaller than the probable maximum.

3. A probable maximum snowmelt flood in conjunction with a major rainstorm less severe than the probable maximum storm for that season.

All of the Bureau reservoirs are designed to accommodate an IDF (Inflow Design Flood) and an MDE (Maximum Design Earthquake). The IDF and the MDE are defined as the flood and the earthquake, respectively, which control the design of a specific dam and its related features.

The evaluation of the protection level is essential for formulating alternatives to solve the problem. This evaluation will result in one of three general cases from which to select loading conditions.

Case A—Maximum Loading Conditions

This would be the case where the level and proximity of the downstream hazard make it clear at the outset of the problem that the consequences of dam failure in terms of potential loss of life or property damage would be unacceptable regardless of how remote the chance of failure may be. Thus, the loading conditions for the various alternatives are established at the maximum level (MCE, PMF, etc.).

Case B—Loading Conditions Determined by Economic Analysis

This would be the case where the level and/or remoteness of the downstream hazard are such that it is apparent (or becomes apparent) that incremental impact of dam failure would not significantly change the potential for loss of life or other nonmonetary factors, and that an economic analysis in which the costs and benefits of reducing the hazard becomes the primary consideration.

Case C—Loading Conditions as a Parameter in the Ultimate Decision Making Process

This case is one where the incremental consequences of dam failure (with or without consideration of warning or other nonstructural modifications) do not clearly indicate that the dam falls under Case A or Case B. Comparison of alternatives for this case would include the economic comparison as for Case B, but would require a more comprehensive assessment of the incremental effects of dam failure on potential for loss of life (with and without warning system) as well as the incremental effects socially, environmentally, and politically for each alternative and load level.

Additional Considerations for Existing Dams

It is desirable that existing dams meet the Bureau's basic IDF criteria for proposed dams. Therefore, a reevaluation of an existing dam with respect to selecting and accommodating the IDF should be based on the same basic criteria. The reevaluation should be performed in a systematic manner taking into account present conditions at the dam, reservoir, and downstream flood plain. Present or anticipated conditions may reduce or increase requirements related to selection and accommodation of the IDF. Performance information for the dam and operation history of the reservoir may reduce uncertainties that were conservatively accounted for in the original design. Likewise, land use pattern around the reservoir rim and downstream from the dam may now be well established. It is recognized that for some existing dams where hazardous conditions prevail, there is the potential, if accomplished in a very cautious manner, for selection of an IDF of lesser magnitude than the PMF; this may be justified because of irreconcilable conditions that have developed since construction. However, any relaxation of established criteria is undertaken with extreme caution on a case-by-case basis after the consequences of dam failure have been evaluated and quantified.

Federal Energy Regulatory Commission (FERC)
(From letter dated June 12, 1984)

The following is extracted from material submitted by FERC:

The criteria presented herein apply to both the review of designs by Commission staff prior to licensing and review of licensed projects by independent consultants under Part 12 of the Commission's regulations.

The adequacy of new and existing projects for extreme flood conditions is evaluated by considering the hazard potential which would result from failure of the project works during flood flows. If structural failure would present a hazard to human life or cause significant property damage, the project is evaluated as to its ability to withstand the loading or overtopping which may occur from a flood up to the probable maximum. If structural failure would not present a hazard to human life or cause significant property damage, a spillway design flood of lesser magnitude than the probable maximum flood would be acceptable provided that the basis for the finding that structural failure would not present a hazard to human life is significantly documented. As a result of the publications of Hydrometeorological Reports Nos. 51 (Schreiner and Riedel, 1978) and 52 (Hansen et al., 1982), the Commission staff has adopted guidelines [shown below] for evaluating the spillway adequacy of all licensed and exempted projects located east of the 105th meridian.

(1) For existing structures where a reasonable determination of the Probable Maximum Precipitation (PMP) has not previously been made using suitable methods and data such as contained in HMR No. 33 (Riedel et al., 1956) or derived from specific meteorologic studies, or the PMF has not been properly determined, the ability of the project structures to withstand the loading or overtopping which may occur from the PMF must be reevaluated using HMR Nos. 51 and 52.

(2) For existing structures where a reasonable determination of the PMP has previously been made, a PMF has been properly determined, and the project structures can withstand the loading or overtopping imposed by that PMF, the reevaluation of the adequacy of the spillway using HMR Nos. 51 and 52 is not required. Generally no PMF studies will be repeated solely because of the publication of HMR Nos. 51 and 52. However, there is no objection to using the two reports for necessary PMF studies for any water retaining structure.

(3) For all unconstructed projects and for those projects where any proposed or required modification will significantly affect the stability of water impounding project structures, the adequacy of the project spillway must be evaluated using: (a) HMR Nos. 51 and 52, or (b) specific basin studies where the project lies in the stippled areas on Figures 18 through 47 of HMR No. 51.

Forest Service, U.S. Department of Agriculture
(From letter dated May 23, 1984)

The following is extracted from material submitted by the Forest Service:

Hazard-Potential Assessment

The hazard class (see Definitions) is based on the potential damage that can be anticipated in the event of dam failure. Potential damage is to be assessed under clear weather conditions with normal base inflow to the reservoir and the water surface at the elevation of the uncontrolled spillway crest.

Hydrologic Criteria

Select a spillway design flood based on an evaluation of the potential risk and consequences of flooding under both present and future conditions. The flood selected for design of spillways should have virtually no chance of being exceeded when failure could pose a hazard to life or cause significant property damage. The spillway capacity and/or storage capacity shall safely handle the design flood without failure.

Where a spillway design flood range is shown in Table A-1, select the magnitude commensurate with the involved risk.

It is recognized that failure of some dams with a relatively small reservoir capacity may have little influence on the potential damage anticipated during the spillway design flood event.

Exceptions to the recommended spillway design flood magnitude may be permissible for some structures. Requests for an exception must include sufficient documentation to demonstrate that economic loss and/or the po-

TABLE A-1 Recommended Spillway Design Flood

Hazard Potential	Size Class	Spillway Design Flood
High	A	PMF
	B	PMF
	C	1/2 PMF to PMF
	D	100 yr to 1/2 PMF
Moderate	A	PMF
	B	1/2 PMF to PMF
	C	100 yr to 1/2 PMF
Low	A	1/2 PMF to PMF
	B	100 yr to 1/2 PMF
	C	50 yr to 100 yr

tential for loss of life resulting from dam failure during occurrence of the proposed spillway design flood would be essentially the same as would occur without a dam failure. The Regional Director of Engineering must approve exceptions to the recommended spillway design flood. When documentation is not available to support an exception, use the recommended spillway design flood criteria shown in Table A-1.

Definitions

1. *Administrative.* The classification of a project for administrative purposes, based on height and storage.

    a. *Class A Projects.* Dams that are 100 feet high or more, or that impound 50,000 acre-feet or more of water.

    b. *Class B Projects.* Dams that are 40 to 99 feet high, or that impound 1,000 to 49,999 acre-feet of water.

    c. *Class C Projects.* Dams that are 25 to 39 feet high, or that impound 50 to 999 acre-feet of water.

    d. *Class D Projects.* Dams that are less than 25 feet high and that impound less than 50 acre-feet of water. The inclusion of structures less than 6 feet high or impounding less than 15 acre-feet of water is optional with the approving officer.

2. *Hazard Potential.* The classification of a dam based on the potential for loss of life or damage in the event of a structural failure under clear weather conditions with normal base inflow to the reservoir and the water surface at the elevation of the uncontrolled spillway crest.

    a. *Low Hazard.* Dams built in undeveloped areas where failure would result in minor economic loss, damage would be limited to undeveloped or agricultural lands, and improvements are not planned in the forseeable future. Loss of life would be unlikely.

b. *Moderate Hazard.* Dams built in areas where failure would result in appreciable economic loss, with damage limited to improvements, such as commercial and industrial structures, public utilities, and transportation systems, and serious environmental damage. No urban development and no more than a small number of habitable structures are involved. Loss of life would be unlikely.

c. *High Hazard.* Dams built in areas where failure would likely result in loss of life or where economic loss would be excessive; generally, areas or urban- or community-type developments that have more than a small number of habitable structures.

Interagency Committee on Dam Safety (ICODS)
(From draft of proposed "Federal Guidelines for Selecting and Accommodating Inflow Design Floods for Dams" prepared by a working group and submitted to the Chairman of ICODS by letter dated October 11, 1983)

The following is extracted from the draft guidelines:

Selecting an IDF for the hydrologic safety design of a dam requires balancing the likelihood of failure by overtopping against the consequences of dam failure. Consequences of failure include the loss of life and social, environmental, and economic impacts. The inability to accurately define flood probabilities for rare events, and to accurately assess the potential loss of life and economic impact of failure when it would occur, dictate use of procedures which provide some latitude to meet site-specific conditions in selecting the IDF.

The PMF should be adopted as the IDF in those situations where consequences attributable to dam failure from overtopping are unacceptable. The determination of unacceptability exists when the area affected is evaluated and factors indicate loss of human life, extensive property and environmental damage, or serious social impact may be expected as a result of dam failure.

A flood less than the PMF may be adopted as the IDF in those situations where the consequences of dam failure are acceptable. Acceptable consequences exist when evaluation of the area affected and factors in section F.1.c. [which material relates to evaluating impacts of dam failure] show one of the following conditions:

- There are no permanent human habitations, or commercial or industrial development, nor are such habitations, or commercial or industrial developments projected to occur within the potential hazard area in the foreseeable future and transient population is not expected to be affected.

- There are only a few permanent human habitations within the potential hazard area that would be impacted by failure of the dam and there would be no significant increase in the hazard resulting from the occurrence of floods larger than the proposed IDF up to the PMF. An example is where impoundment storage is small and failure would not add appreciable volume to the outflow hydrograph, and, consequently, the downstream inundation would be essentially the same with or without failure of the dam. The consequences of dam failure would not be acceptable if the hazard to these habitations was increased appreciably by the failure flood wave or level of inundation, e.g., the case where failure of a storage reservoir would add appreciably to the outflow hydrograph.

In addition to the conditions listed in section F.1.c. [which material relates to evaluating impacts of dam failure], the selected magnitude of the IDF should be based on the following special considerations:

- Dams which provide vital community services such as municipal water supply or energy may require a high degree of protection against failure to ensure those services are continued during and following extreme flood conditions when alternate services are unavailable.
- Dams should be designed to not less than some minimum standard to reduce the risk of loss of benefits during the life of the project; to hold O&M costs to a reasonable level; to maintain public confidence in agencies responsible for dam design, construction, and operation; and to be in compliance with local, State, or other regulations applicable to the facility.

National Weather Service (NWS), National Oceanic and Atmospheric Administration, U.S. Department of Commerce
(From letter dated June 1, 1984)

The following is extracted from material submitted by the NWS:

Although the agency is not directly involved with dams and design criteria for dams, the National Weather Service has furnished extensive material on Probable Maximum Precipitation estimates and the techniques for developing such estimates, which provide the bases for the most conservative criteria for spillway design. The PMP has been defined as "the theoretically greatest depth of precipitation for a given duration that is physically possible over a given size storm area at a particular geographical location at a certain time of year."

From this definition, theoretically the PMP has zero probability of actual occurrence. A report (Riedel, J. T., and Shreiner, L. C. 1980) compares the greatest known storm rainfall depths with generalized PMP estimates for the

United States east of the 105th meridian and west of the Continental Divide. This was done for rainfall depths averaged over six area sizes (10, 200, 1000, 5000, 10,000, and 20,000 mi²) each for five durations (6, 12, 24, 48, and 72 hr) covering the eastern United States. This gives comparisons for 30 combinations of area sizes and durations. The western states comparisons are more difficult to make, so only six combinations were made. These combinations were: for 10 mi² and durations of 6 and 24 hours; for 500 mi² and durations of 24 and 48 hours; and for 1000 mi² and durations of 24 and 48 hours.

For the eastern United States there were the following number of incidents (from the 30 combinations of area size and duration) where the rainfall was within the indicated percent of the PMP:

Percent of PMP equaled or exceeded	70	80	90
No. of incidents	160	49	4

For the western states from only six combinations of area size and duration the number of incidents were:

Percent of PMP equaled or exceeded	70	80	90
No. of incidents	16	5	0

Another comparison shows that for the eastern states there were 170 separate storms which had depths exceeding 50% of PMP for at least one area size and duration. The comparable number for the western states is 66. It should be noted that both the number of storms and storm incidents are directly related to the number of area and duration combinations compared.

Soil Conservation Service, U.S. Department of Agriculture

(From letter dated May 21, 1984, Criteria presented in Technical Release No. 60, "Earth Dams and Reservoirs," revised August 1981)

SCS has established three classes of dams as follows:

Class (a)   Dams located in rural or agricultural areas where failure may damage farm buildings, agricultural land, or township and country roads.

Class (b)   Dams located in predominantly rural or agricultural areas where failure may damage isolated homes, main highways or minor railroads or cause interruption of use or service of relatively important public utilities.

Class (c)   Dams located where failure may cause loss of life, serious damage to homes, industrial and commercial buildings, important public utilities, main highways, or railroads.

Minimum criteria for spillway design to prevent overtopping of dams are given in Table 2-5 of Technical Release No. 60, which is reproduced below. Much less demanding criteria for small low hazard potential dams (having effective heights of 35 feet or less and for which the product of the storage in acre-feet times the effective height in feet is less than 3,000) are given in SCS Practice Standard 378 for Ponds which provides for minimum spillway design storms having 10-year to 50-year frequencies.

SCS does not differentiate between new dams and existing dams in its criteria.

Tennessee Valley Authority
(From letter dated June 7, 1984)

TVA uses a hazard classification for structures described as follows:

High Hazard

The high hazard classification includes structures whose failure during floods would likely cause serious social or economic loss. Unless specific studies show otherwise, structures 100 feet or more in height or with 50,000 acre-feet or more of total capacity at maximum flood levels shall arbitrarily be classified as high hazard.

Medium Hazard

The medium hazard classification includes structures whose failure during floods would cause significant but not serious social or economic loss. When a higher hazard situation is not evident, structures over 25 feet but less than 100 feet in height or with total capacity at maximum flood levels greater than 5,000 acre-feet but less than 50,000 acre-feet shall arbitrarily be classified as medium hazard unless specific studies show otherwise.

Low Hazard

The low hazard classification includes any structure whose failure during floods would likely cause only minor social or economic loss. Structures not in the high hazard or medium hazard classifications defined above shall be classified as low hazard when neither existing nor prospective future conditions indicate that a higher hazard situation is to be expected.

TVA's guidelines provide that high hazard structures will be tested with the *probable maximum flood*, medium hazard structures with the *TVA maximum probable flood* and low hazard structures with a design flood "appropriate to the economic life and planned purpose of the structure." The *probable maximum flood* and the *TVA maximum probable flood* determinations are to be based upon combinations of hydrologic factors which are selected to prevent unrealistic combinations of hydrologic conditions.

TABLE 2-5 of T.R. No. 60: Minimum Emergency Spillway Hydrologic Criteria

Class of Dam	Product of Storage × Effective Height	Existing or Planned Upstream Dams	Precipitation Data for[1] Emergency Spillway Hydrograph	Freeboard Hydrograph
(a)[2]	Less than 30,000	None	P_{100}	$P_{100} + 0.12 \, (PMP - P_{100})$
	Greater than 30,000	None	$P_{100} + 0.06 \, (PMP - P_{100})$	$P_{100} + 0.26 \, (PMP - P_{100})$
	All	Any[3]	$P_{100} + 0.12 \, (PMP - P_{100})$	$P_{100} + 0.40 \, (PMP - P_{100})$
(b)	All	None or any	$P_{100} + 0.12 \, (PMP - P_{100})$	$P_{100} + 0.40 \, (PMP - P_{100})$
(c)	All	None or any	$P_{100} + 0.26 \, (PMP - P_{100})$	PMP

[1]P_{100}—Precipitation for 100-year return period. PMP—Probable maximum precipitation.
[2]Dams involving industrial or municipal water are to use minimum criteria equivalent to that of class (b).
[3]Applies when the upstream dam is located so that its failure could endanger the lower dam.

Probable Maximum Floods are based on probable maximum precipitation (PMP) defined as "that rainfall over a particular basin which has vitually no risk of being exceeded." PMP estimates come "from National Weather Service studies applicable to watersheds in the Tennessee Valley."

TVA Maximum Probable Floods used for the design of the older TVA dams were based on maximum observed runoff rate diagrams and maximum observed storms. Currently such floods are based upon TVA maximum probable precipitation estimates defined "as that magnitude of rainfall over a particular basin which is equivalent to maximum storms that have been observed within regions of similar meteorological character. Storm rainfall amounts are based upon 'TVA precipitation' from the National Weather Service studies applicable to watershed in the Tennessee Valley."

Guidelines for other factors affecting the development of probable maximum and TVA maximum probable floods generally call for average or median conditions observed during maximum past floods.

A "decision tree," illustrating the TVA approach to hydrologic design is shown in Figure A-1.

U.S. Army Corps of Engineers (For Corps Projects)
(From letter dated May 24, 1984)
(Spillway and freeboard criteria are from Engineer Circular 1110-2-27,
dated August 1, 1966)

The following is extracted from material submitted by the Corps:

The Corps has established four functional design standards for new dams designed by the Corps as follows:

Standard 1:   Design dam and spillway large enough to assure that the dam will not be overtopped by floods up to probable maximum categories.

Standard 2:   Design the dam and appurtenances so that the structure can be overtopped without failing and, insofar as practicable, without suffering serious damage.

Standard 3:   Design the dam and appurtenances in such manner as to assure that breaching of the structure from overtopping would occur at a relatively gradual rate, such that the rate and magnitude of increases in flood stages downstream would be within acceptable limits, and that damage to the dam itself would be located where it could be most economically repaired.

Standard 4:   Keep the dam low enough and storage impoundments small enough that no serious hazard would exist downstream in the event of

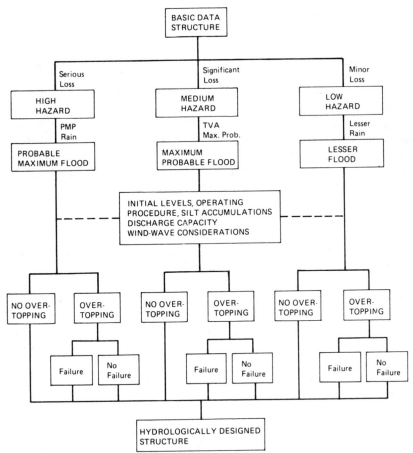

FIGURE A-1 Decision tree of TVA.

breaching, and so that repairs to the dam would be relatively inexpensive and simple to accomplish.

In application of these standards most Corps of Engineers new dams are designed to pass the PMF with full freeboard; the exceptions being run-of-river developments, diversion dams, and small dams with small impounding capacities and low downstream hazard potentials.

The Corps uses the same hydrologic safety criteria in design of new dams and in analyzing and upgrading existing dams. However, because of the limited resources available at any one time, the Corps uses a decision tree involving consideration of relative existing project capabilities in setting priorities for such upgrading.

U.S. Army Corps of Engineers—(for National Dam Inspection Program)
(From ER 1110-2-106 and ETL 110-2-234)

The following is extracted from material submitted by the Corps:

In cooperation with other federal agencies, state agencies and other groups and individuals knowledgeable about dam safety matters, the Corps of Engineers developed a pamphlet, "Recommended Guidelines for the Safety Inspection of Dams," to guide the inspection of nonfederal dams authorized in 1972 by P.L. 92-367.

The Corps' recommended guidelines provide for classifying dams by height of dam and storage impounded and, also, by hazard potentials in the downstream areas in the event of failure of the dams. These provisions (as modified by a change in September 1979) are described in the following:

Size.   The classification for size based on the height of the dam and storage capacity should be in accordance with Table A-2. The height of the dam is established with respect to the maximum storage potential measured from the natural bed of the stream or watercourse at the downstream toe of the barrier, or if it is not across a stream or watercourse, the height from the lowest elevation of the outside limit of the barrier, to the maximum water storage elevation. For the purpose of determining project size, the maximum storage elevation may be considered equal to the top of dam elevation. Size classification may be determined by either storage or height, whichever gives the larger size category.

Hazard Potential.   The classification for potential hazards should be in accordance with Table A-3. The hazards pertain to potential loss of human life or property damage in the area downstream of the dam in event of failure or misoperation of the dam or appurtenant facilities. Dams conforming to criteria for the low hazard potential category generally will be located in rural or agricultural areas where failure may damage farm buildings, limited agricultural land, or township and country roads. Significant hazard

TABLE **A-2**   Size Classification

Category	Impoundment Storage (ac-ft)	Height (ft)
Small	1,000 and 50	40 and 25
Intermediate	1,000 and 50,000	40 and 100
Large	50,000	100

TABLE A-3   Hazard Potential Classification

Category	Loss of Life (extent of development)	Economic Loss
Low	None expected (no permanent structures for human habitation)	Minimal (undeveloped to occasional structures or agriculture)
Significant	Few (no urban developments and no more than a small number of inhabitable structures)	Appreciable (notable agriculture, industry, or structures)
High	More than few	Excessive (extensive community, industry, or agriculture)

potential category structures will be those located in predominantly rural or agricultural areas where failure may damage isolated homes, secondary highways or minor railroads or cause interruption of use or service of relatively important public utilities. Dams in the high hazard potential category will be those located where failure may cause serious damage to homes, extensive agricultural, industrial and commercial facilities, important public utilities, main highways, or railroads.

The Corps issued the following supplementary guidelines regarding classifying dams as unsafe:

A finding that a dam will not safely pass the flood indicated in the Recommended Guidelines does not necessarily indicate that the dam should be classified as unsafe. The degree of inadequacy of the spillway to pass the appropriate flood and the probable adverse impacts of dam failure because of overtopping must be considered in making such classification. The following criteria have been selected which indicate when spillway capacity is so seriously inadequate that a project must be classified as unsafe. All of the following conditions must prevail before designating a dam unsafe:

a. There is high hazard to loss of life from large flows downstream of the dam.

b. Dam failure resulting from overtopping would significantly increase the hazard to loss of life downstream from the dam from that which would exist just before overtopping failure.

c. The spillway is not capable of passing one-half of the probable maximum flood without overtopping the dam and causing failure.

TABLE A-4   Hydrologic Evaluation Guidelines: Recommended Spillway
Design Floods

Hazard	Size	Spillway Design Flood (SDF)[a]
Low	Small	50- to 100-yr frequency
	Intermediate	100-yr to ½ PMF
	Large	½ PMF to PMF
Significant	Small	100-yr to ½ PMF
	Intermediate	½ PMF to PMF
	Large	PMF
High	Small	½ PMF to PMF
	Intermediate	PMF
	Large	PMF

[a]The recommended design floods in this column represent the magnitude of the spillway design flood (SDF), which is intended to represent the largest flood that need be considered in the evaluation of a given project, regardless of whether a spillway is provided; i.e., a given project should be capable of safely passing the appropriate SDF. Where a range of SDF is indicated, the magnitude that most closely relates to the involved risk should be selected.

100-yr = 100-Year Exceedance Interval. The flood magnitude expected to be exceeded, on the average, once in 100 years. It may also be expressed as an exceedance frequency with a one percent chance of being exceeded in any given year.

PMF = Probable Maximum Flood. The flood that may be expected from the most severe combination of critical meteorologic and hydrologic conditions that are reasonably possible in the region. The PMF is derived from probable maximum precipitation (PMP), which information is generally available from the National Weather Service, NOAA. Most federal agencies apply reduction factors to the PMP when appropriate. Reductions may be applied because rainfall isohyetals are unlikely to conform to the exact shape of the drainage basin and/or the storm is not likely to center exactly over the drainage basin. In some cases local topography will cause changes from the generalized PMP values; therefore, it may be advisable to contact federal construction agencies to obtain the prevailing practice in specific areas.

U.S. Nuclear Regulatory Commission (NRC)
(From letter dated June 8, 1984)

The following is extracted from material furnished by NRC:

Although the Nuclear Regulatory Commission (NRC) by itself does not plan, design, construct or operate dams, the NRC does regulate dams whose failure could result in a radiological risk to public health and safety. By virtue of this regulatory responsibility, which is described in the Code of Federal Regulations, the NRC has developed guidelines and design criteria for addressing flood and earthquake hazards, which applicants for permits and licenses to operate nuclear facilities are required to meet.

The regulations and criteria are primarily related to the design and construction of nuclear power plant structures, systems, and components. The Nuclear Regulatory Commission is also involved with the regulation of embankment retention systems for uranium mill tailings where the radiological risk to the public health and safety is considerably less than it is with nuclear power plants. In recognition of this reduced risk, less stringent flooding and earthquake design criteria have been considered for special site conditions (small dams built in isolated areas), where the dam failure would neither jeopardize human life nor create damage to property or the environment beyond the sponsor's legal liabilities and financial capabilities.

Nuclear power plants should be designed to prevent the loss of capability for cold shutdown and maintenance thereof resulting from the most severe flood conditions that can reasonably be predicted to occur at a site as a result of severe hydrometeorological conditions, seismic activity, or both.

The conditions resulting from the worst site-related flood probable at the nuclear power plant (e.g., PMF, seismically induced flood, seiche, surge, severe local precipitation) with attendant wind-generated wave activity constitute the design basis flood conditions that safety-related structures, systems, and components are designed to withstand.

There will always be some catchment area contributing runoff into the [uranium mill] tailing retention system. This may vary from the area of the system itself to a substantial area incorporating the drainage area of streams entering the valley across which a retention dam is constructed. Substantial runoff volumes and flows can result from heavy precipitation or snowmelt over relatively small catchment areas.

The maximum runoff used in the designing is usually called the Spillway Design Flood (SDF), representing the largest flood that needs to be analyzed, regardless of whether or not a spillway is provided. The magnitude of the SDF (flood volume, peak, flow, etc.) as adopted in the United States for the past 30 years is equal to that of the Probable Maximum Flood at the site of the dam.

For smaller retention dams built on isolated streams in areas where failure would neither jeopardize human life nor create damage to property or the environment beyond the sponsor's legal liabilities and financial capabilities, less conservative flood design criteria may be used in the design. However, the selection of the design flood needs to be at least compatible with the guidelines set for the by the Corps of Engineers ["Recommended Guidelines for Safety Inspection of Dams"].

If decant or other reclaim systems have not been designed specifically to pass the design flood, other measures need to be taken. Those other measures may be one or a combination of the following:

a. Storing the whole volume of flood runoff. Sufficient freeboard should always be available to provide the necessary storage capacity without over-topping the dam.

b. Providing a spillway or diversion channels to convey runoff water safely past the dam.

Because of the toxic nature of the impounded material, *a.* is preferred.

PART 2—STATE AGENCIES RESPONSIBLE FOR DAM SAFETY

Alaska
(From letter dated May 16, 1984)

Alaska is now in the process of preparing legislation and developing criteria for review of plans for dams and inspections and relies heavily on Corps of Engineers criteria. Hydrologic evaluation criteria for spillways currently used are as follows:

Height of dam (ft) or Volume impounded (A.F.)	10' to 40' 50 to 1000	40' to 100' 1,000 to 50,000	100' 50,000
Low hazard	50-100 yr freq.	100 yr to $1/2$ PMF	$1/2$ PMF to PMF
Significant hazard	100 yr to $1/2$ PMF	$1/2$ PMF to PMF	PMF
High hazard	$1/2$ PMF to PMF	PMF	PMF

Arizona
(From letter dated June 7, 1984)

The Arizona Department of Water Resources is completing a revision of its "Guidelines for the Determination of Spillway Capacity Requirements." Extracts from the revised draft guidelines, which have been in use for some time, are shown in Table A-5.

Size Classification

Dams are classified into small, medium and large sizes. A numerical rating procedure, based on the descriptive characteristics of height and

TABLE A-5 Hazard Potential Classification

Category	Loss of Life (extent of development)	Economic Loss
Low	None expected (no permanent structures for human habitation)	Minimal (undeveloped to occasional structures or agriculture)
Significant	Few (no urban developments and no more than a small number of inhabitable structures)	Appreciable (notable agriculture, industry, or other structures)
High	More than few	Excessive (extensive community, industry, or agriculture)

reservoir capacity, has been developed to determine the dam size classification.

Height is measured from the lowest elevation of the outside limit of the dam (usually the downstream toe) to the spillway crest, or top of spillway gates if so equipped. For dams with no spillway, the height is measured to the crest of the dam.

Capacity, in acre-feet, is measured to the spillway crest or top of the spillway gates, if so equipped. For dams with no spillway, capacity is measured to the dam crest.

The categories and corresponding rating factors are shown in Table A-6. A numerical rating is computed for each dam by adding the corresponding rating factors for each of the two categories. For example, a dam that is 65 feet in height and has a reservoir capacity of 22,000 acre-feet would have a rating of $(3+4=7)$.

Small dams have a rating in the range 0-2, medium dams in the range 3-7, and large dams, 8 or greater.

All new dams, existing dams that are being enlarged or improved, and dams being reevaluated for safety may have spillways of lesser capacity than that outlined by Table A-7 as discussed below.

A spillway capacity less than outlined above will be acceptable, where the owner (or his engineer) can demonstrate to the department that the incremental damages due to failure of the dam are insignificant and will not cause loss of life. The analysis shall be based upon the dam failure caused by a flood that just exceeds the routing capacity of the reservoir. The result shall be compared to the pre-failure conditions such as the spillway discharge and any reasonable rainfall runoff occurring between the dam site and the point(s) of interest below the dam. The burden of proof rests with the owner.

TABLE A-6   Size Classification

Category	Rating Factor	Category	Rating Factor
Height (ft)		Reservoir Capacity (ac-ft)	
6-24	0	15-499	0
25-39	1	500-999	1
40-59	2	1,000-2,999	2
60-79	3	3,000-9,999	3
80-99	4	10,000-24,999	4
100+	5	25,000+	5

TABLE A-7   Spillway Capacity Requirements: Recommended Spillway Design Floods

Hazard Category	Size Designation	Inflow Design Flood Magnitude
Low	Small	100 yr
	Medium	100 yr to $1/2$ PMF
	Large	$1/2$ PMF
Significant	Small	100 yr to $1/2$ PMF
	Medium	$1/2$ PMF
	Large	$1/2$ PMF to PMF
High	Small	$1/2$ PMF
	Medium	$1/2$ PMF to PMF
	Large	PMF

Arkansas
(From letter dated May 14, 1984)

The following is extracted from "Rules Governing the Arkansas Dam Safety Program":

The spillway capacity must be capable of passing the spillway design flood (SDF) without endangering the safety of the dams. The spillway design must include sufficient capacity and freeboard to prevent overtopping of the dam, have sufficient strength to prevent structural failure, and an adequate energy-dissipating device at the outlet. The following minimums will apply (PMF = probable maximum flood—The flood that may be expected from the most severe combination of critical meteorologic and hydrologic conditions that are reasonably possible in the region.):

Equal to or Greater Than

Class	Storage Capacity	Drainage Area	Spillway Design Flood
1	10,000 (ac-ft)	10 sq. mi.	0.75 PMF
2	5,000 (ac-ft)	1000 ac.	0.50 PMF
3	1,000 (ac-ft)	100 ac.	0.25 PMF
4	20 (ac-ft)	0	100 year

California
(From letter dated June 1, 1984)

The following is quoted from a letter from the Chief, Division of Safety of Dams:

In response to your May 4, 1984 letter, we are outlining our approach to hydrologic and earthquake related safety criteria and standards. As a matter of policy we do not publish standards or criteria so the information provided herein has been compiled from several internal documents specifically to answer your letter.

Hydrology—spillway capacity. The basic requirement is stated: "The size and type of dam and its vulnerability to failure because of an inadequate spillway shall be considered in the selection of the magnitude of the spillway design flood, and consequently the spillway capacity."

The minimum design flood required is a one in 1000 year flood and the maximum is a probable maximum flood as derived from the probable maximum precipitation determined from Hydrometeorological Report No. 36 (U.S. Weather Bureau, 1961a) or Technical Paper No. 38 (U.S. Weather Bureau, 1960) as appropriate for the drainage area. The return period for the flood is selected by using a rating system that considers (1) the reservoir capacity, (2) dam height, (3) estimated number of people that would have to be evacuated in anticipation of dam failure, and (4) potential downstream damage. The system is such that only remote farm dams qualify for the one in 1000 year floods. Typically probable maximum floods are required for dams that impound 1000 acre feet or more, are at least 50 feet high, the estimated evacuation is at least 1000 people, and the damage potential is $20,000,000 or greater. The scale for floods between the 1000 year and probable maximum is continuous; so a dam with a slightly lower rating in one of the four factors than the example would require a statistical flood equal to about 90 percent of a probable maximum flood.

New embankment dams must pass the spillway design with a minimum of 1½ feet of residual freeboard above the maximum reservoir stage. Additional freeboard is required for severe wave conditions. Residual freeboard for new concrete dams is based on the ability of abutment and foundation to resist damage from overpour. Existing dams must only safely pass the spillway design flood.

The Department of Water Resources as owner of 16 dams uses probable maximum floods for all dams "except for low diversion dams and other small dams impounding relatively insignificant quantities as compared to the volume of flood flows." These dams must also conform to the Division of Safety of Dam requirements.

Colorado
(From letter dated June 6, 1984)

Colorado's spillway capacity criteria have been stated as follows:

The inflow design flood for a reservoir [generally] is the probable maximum flood. However, it may be smaller than a probable maximum flood provided it can be shown that the incremental damages due to failure of the dam are insignificant and will not cause loss of life.

The analysis shall be based upon the dam failure caused by a flood which just exceeds the routing capacity of the reservoir. This result shall be compared to the pre-failure conditions such as the spillway discharge and any reasonable rainfall runoff occurring between the dam site and the point(s) of interest below the dam.

A minor dam situated in a remote area, where loss of life or property damage is not envisioned, will not require an incremental damage analysis. However, the minimum size spillway must safely pass the 100-year flood.

In order to ensure the safety of the dam embankment during the IDF, all new dams, and enlargements of existing dams, shall have spillways which can safely pass the inflow design flood and have a minimum of one foot of residual freeboard exclusive of camber. Existing dams shall be able to pass the IDF safely.

In the design of the spillway, all new dams shall have a minimum normal water level freeboard of not less than five feet and existing dams may have a minimum of three feet of freeboard, if it can be shown that the structure will be safe. Exceptions will be on a case by case basis.

Georgia
(From letter dated June 5, 1984)

Dam safety criteria were specified by the original 1978 Georgia State Dams Act. These criteria have been amended by legislation in 1982 and 1984. Current criteria for spillway capacity are shown below.

*Spillway Requirements**

1) Small Dams (height < 25 ft and max. storage < 500 acre-feet)—25 % of PMP

2) Medium Dams (25 ft < height < 35 ft or 500 acre-feet < storage < 1000 acre-feet)—33 % of PMP

3) Large Dams (35 ft < height < 100 ft or 1000 acre-feet < storage < 50,000 acre-feet)—50 % of PMP

4) Very Large Dams (100 ft < height or 50,000 acre-feet < storage)—100 % of PMP

*Based on visual inspection and detailed hydrologic and hydraulic evaluation, including documentation of competent design and construction procedures, up to a 10 percent lower requirement (22.5, 30, 45, 90) can be accepted, at the discretion of the director, provided the project is in an acceptable state of maintenance. The design storm may also be reduced if the applicant's engineer can successfully demonstrate to the director by engineering analysis that the dam is sufficient to protect against probable loss of human life downstream at a lesser design storm.

Hawaii
(From letter dated May 31, 1984)

State does not have an authorized dam safety program.

Illinois
(From letter dated May 1984)

Hydrologic requirements have been summarized by a state official as follows:

Class I (High Hazard Potential) Dams—All dams in this hazard potential classification shall hold and pass the following floods:

Large Dams—PMF
Intermediate Dams—PMF
Small Dams—1/2 PMF to PMF

Should dams in this hazard potential classification be designed to have emergency spillways, the principal spillway shall pass at least the entire 100-year flood before the emergency spillway functions, unless special site conditions justify variations.

Class II (Moderate Hazard Potential) Dams—All dams in this hazard potential classification shall hold and pass the following floods:

Large Dams—PMF
Intermediate Dams—$1/2$ PMF to PMF
Small Dams—100-yr to $1/2$ PMF

Should dams in this hazard potential classification be designed to have emergency spillways, the principal spillway shall pass at least the entire 50-year flood before the emergency spillway functions, unless special site conditions justify variations.

Class III (Low Hazard Potential) Dams—All dams in this hazard potential classification shall hold and pass the following floods:

Large Dams—$1/2$ PMF to PMF
Intermediate Dams—100 yr to $1/2$ PMF
Small Dams—100 yr

Should dams in this hazard potential classification be designed to have emergency spillways, the principal spillway shall pass the 25-year flood before the emergency spillway functions, unless special site conditions justify variations.

The following is quoted from letter of the Chief, Illinois Bureau of Resource Management:

We are very interested in the study now being conducted by the National Research Council. In fact, we have postponed enforcement actions that involve inadequate spillway capacity related to the probable maximum flood pending completion of your research effort, unless there is a definite, immediate safety hazard. I hope that your evaluation of the various criteria used by the respondents will include a definitive statement of the appropriate standards which are discerned as being reasonable by the National Research Council. Such a statement would provide a positive benefit to the respondents in assessing their own standards and perhaps aid in the development of more uniform standards nationwide.

Indiana
(From pamphlets supplied by Edwin B. Vician)

By a 1945 law the Indiana Flood Control and Water Commission was established. By a 1961 law the Commission was granted authority over dams and authorized to issue rules, regulations and standards for maintenance and operation. However, the data furnished do not include any criteria for dams.

Kansas
(From letter dated June 12, 1984)

The following is extracted from information furnished by the Chief Engineer-Director of the Kansas State Board of Agriculture:

Structures built in Kansas are predominately earthen enbankments located in a rural or semi-rural setting. In order to assist the designer and to provide an acceptable level of consistency of design for this area, we have adopted Engineering Guide Nos. 1 and 2 by reference in our rules and regulations.

These standards and design criteria were based upon many years of experience and we feel they are both adequate and practical for conditions in Kansas.

Table No. 2, page 11, of Engineering Guide No. 1 (reproduced below) outlines the use of variable probable maximum precipitation based upon hazard classification of structure. In addition, please note the required minimum spillway dimensions. We may also require more extensive safety measures to be incorporated into the design if the magnitude and location of the structure warrants consideration beyond perimeters set forth in Table No. 2.

These guidelines provide the general minimum hydrologic requirements of the Division of Water Resources for the design and construction of earthfill dams. They are not intended to constitute a text for design and construction. Final determination of the acceptability of design and adequacy of the plans and specifications will be made on an individual basis.

The following list of definitions relates to the data shown on the Table No. 2 mentioned above:

Effective Height of Dam—the difference in elevation between the crest of the emergency spillway and the original streambed on the centerline of the dam.

Effective Storage—the volume of the reservoir below the crest of the emergency spillway.

Size Factor—the product of the Effective Height of Dam (in feet) and the Effective Storage (in acre-feet).

Size of Dams—

(1) Those dams whose effective height is less than 25 feet; effective storage is less than 50 acre-feet; and size factor is less than 1,250.

(2) Those dams whose effective storage is greater than 50 acre-feet; and size factor is between 1,250 and 3,000.

(3) Those dams whose effective storage is greater than 50 acre-feet; and size factor is between 3,000 and 30,000.

TABLE 2 from Engineering Guide No. 1 (Kansas State Board of Agriculture)

Size	Hazard Class	[1] Minimum Detention Storage	[2] Precipitation for Inflow Hydrograph	Minimum* Freeboard	[4] Minimum Emergency Spillway Dimensions	
					Depth	Width
(1) Eff Storage <50 ac ft Eff Height <25 ft Factor <1,250	a	Volume of runoff from P_2 or 18″ elevation difference [3], whichever is greater	P_{50}	1	3	20
	b		0.25 PMP	2		
	c		0.40 PMP	3		
(2) Eff Storage >50 ac ft Factor: 1,250 to 3,000	a	Volume of runoff from P_2 or 18″ elevation difference [3], whichever is greater	P_{100}	2	3	30
	b		0.25 PMP	2		
	c		0.40 PMP	3		
(3) Eff Storage >50 ac ft Factor: 3,000 to 30,000	a	Volume of runoff from P_2 or 24″ elevation difference [3], whichever is greater	P_{100}	3	5	40
	b		0.30 PMP	3		
	c		0.40 PMP	3		
(4) Eff Storage >50 ac ft Factor >30,000	a	Volume of runoff from P_2 or 30″ elevation difference [3], whichever is greater	0.25 PMP	3	5	40
	b		0.30 PMP	3		
	c		0.40 PMP	3		

[1] Where flood control is one of the primary purposes of a structure, the minimum detention storage shall be as follows:
    (a) Rural protection—detention of runoff from 25-year frequency, 6-hour rainfall, using AMC II.
    (b) Urban protection—detention of runoff from 50-year frequency, 6-hour rainfall, using AMC II.
    In all cases, AMC II will be used in computing minimum detention storage.

[2] In all cases, AMC III will be used in computing runoff for inflow hydrograph.

[3] Elevation difference is vertical distance between crests of principal and emergency spillways.

[4] Structures without emergency spillways will require special consideration.

*DWR may require greater than minimum freeboard dependent on the individual site conditions.

NOTE: AMC II and AMC III refer to antecedent moisture conditions as defined by SCS.

(4) Those dams whose effective storage is greater than 50 acre-feet; and size factor is greater than 30,000.

Hazard Classes of Dams—

Class (a) *Low Hazard*—dams located in rural or agricultural areas where failure may damage farm buildings, limited agricultural land, or county, township and private roads.

Class (b) *Significant Hazard*—dams located in predominantly rural or agricultural areas where failure may endanger few lives, damage isolated homes, secondary highways or minor railroads or cause interruption of use or service of relatively important public utilities.

Class (c) *High Hazard*—dams located in areas where failure may cause extensive loss of life, serious damage to homes, industrial and commercial facilities, important public utilities, main highways or railroads.

Louisiana
(From letter dated May 23, 1984)

State has not yet adopted regulations for dam safety but furnished the following "Excerpt of Proposed Rules and Regulations of Dam Safety Law."

The minimum performance standards for impoundment are as follows:

Hazard	Size	Minimum Spillway Design Flood	Minimum Freeboard
Low	Small	50 to 100 yr	0 foot
	Intermediate	100 yr to $1/2$ PMF	1 foot
	Large	$1/2$ PMF to PMF	3 feet
Significant	Small	100 yr to $1/2$ PMF	0 feet
	Intermediate	$1/2$ PMF to PMF	1 foot
	Large	PMF*	3 feet
High	Small	$1/2$ PMF to PMF	0 foot
	Intermediate	PMF*	1 foot
	Large	PMF*	3 feet

100 yr = 100 year exceedance frequency. The flood magnitude expected to be exceeded, on the average, once in 100 years. It may also be expressed as an exceedance frequency with a one-percent chance of being exceeded in any given year.

PMF = Probable maximum flood. The flood that may be expected from

---

* The primary spillway may be sized to accommodate not less than one-half ($1/2$) of the total PMF, with the remainder of the total PMF accommodated by an emergency spillway.

the most severe combination of critical meteorologic and hydrologic conditions that are reasonably possible in the region. The PMF is derived from probable precipitation, which information is generally available from the following National Weather Service publications:

(1) NOAA Hydrometeorological Report No. 51 (Schreiner and Riedel, 1978) "Probable Maximum Precipitation Estimates—U.S. East of the 105th Meridian".

(2) NOAA Hydrometeorological Report No. 52 (Hansen et al., 1982) "Application of Probable Maximum Precipitation Estimates—U.S. East of the 105th Meridian".

Maine
(From letter dated May 10, 1984)

Regulations are to be developed in near future.

Michigan

State has not issued criteria for dams but follows what is considered good engineering practice. The following is quoted from a statement by the Chief, Dam Safety and Lake Engineering, in the Water Management Division of the State:

The geology of Michigan does not generally allow the construction of large dams as defined by USCOLD. The general criteria for spillway capacity is the 1% frequency flood in rural and undeveloped areas. In urban areas the 0.5% or 200-year frequency flood is a requirement. In addition to this spillway capacity, a freeboard of 1½ foot is required above the design flood elevation for earthen embankment dams. The normal side slopes are 3 to 1 horizontal to vertical for upstream and 2½ to 1 on the downstream. In addition, an appropriate crest width is required. If the side slopes are steeper, a slope analysis by the consulting engineer must be provided to this office for review and acceptance. The foregoing are general criteria for earthen embankment dams which are the majority of the cases in this state. For the larger dams with a higher hazard potential, the guidelines of the National Dam Safety Inspection Program would apply.

Mississippi
(From letter dated May 23, 1984)

The State Dam Safety Coordinator has stated: "We basically follow the criteria and standards of the Soil Conservation Service as published in their

Technical Release No. 60—Earth Dams and Reservoirs. For new dams classified as high hazard we require that they pass the full PMF. For existing high hazard dams we require them to pass 50% PMF."

Missouri
(From letter dated May 10, 1984)

Missouri is in the process of revising its regulations for dam safety. The proposed requirements for spillway design floods are shown [in Table A-8].

This proposed revision specifies use of Probable Maximum Precipitation (PMP) where present regulations refer to Probable Maximum Flood (PMF). The "Environmental Class" listings in this table refer to developments in the area downstream from the dam that would be affected by inundations in the event of dam failure. The classes are defined as:

Class I   Contains 10 or more permanent dwellings or one or more public buildings.

Class II   Contains 1 to 9 permanent dwellings; or one or more camp-

TABLE A-8 Proposed Required Spillway Design Flood Precipitation Values (Missouri)

Dam Type	Stage of Construction	Special Conditions	Environmental Class		
			I	II	III
Conventional or industrial	Completed	Two or more dams in a series	0.75 PMP[a]	0.5 PMP[a]	0.5 PMP[a]
		Storage × height greater than 30,000	0.75 PMP[a]	0.5 PMP[a]	0.4 PMP[a]
		Storage × height less than 30,000[b]	0.75 PMP[a]	0.5 PMP[a]	0.3 PMP[a]
Industrial	Starter	Any	0.5 PMP[a]	0.2 PMP[a]	0.1 PMP[a]
	After starter dam is finished and before final dam is completed	Any	0.75 PMP[a]	0.5 PMP[a]	0.2 PMP[a]

[a]PMP is probable maximum precipitation.
[b]Storage in acre-feet measured at emergency spillway crest elevation and height in feet.

grounds with permanent utility service; or one or more roads with average daily traffic volume of 300 or more; or one or more industrial buildings.

Class III   Everything else.

Nebraska
(From letter dated May 30, 1984)

The following is extracted from letter of Chief, Engineering Branch, Department of Water Resources:

Nebraska Department of Water Resources is principally a regulatory agency of the water resources in the State of Nebraska.

The hydraulic and earthquake criteria acceptable during reviewing plans and specifications of dams are relatively the same as those used in this region by the Federal Agencies, particularly the Soil Conservation Service, Corps of Engineers and Bureau of Reclamation. Occasionally, some deviations of criteria are necessary based on existing site conditions and these are resolved on a site by site basis.

New Jersey
(From letter dated May 25, 1984)

Since January 1978, the state has been using criteria established for U.S. Army Corps of Engineers National Dam Inspection Program, but has now drafted proposed state dam safety regulations. That draft proposes the following criteria for spillway design:

Hazard Classification of Dam	Minimum Spillway Design Flood (SDF)[b]
I-High hazard potential	PMP
II-Significant hazard potential	$1/2$ PMP
III-Low hazard potential	24-hour, 100-year frequency, Type II storm
IV-Small dams[a]	24-hour, 100-year frequency, Type II storm plus 50%

[a]Less than 10 feet high, impounding less than 15 acre/feet, having drainage area of 100 acres or less and not in hazard classification I or II.

[b]Reference to Type II storm is not explained in draft.

New Mexico
(From letter dated June 4, 1984)

The following is extracted from a summary of current practices furnished by the State Engineer:

The State Engineer has not developed a manual of rules and regulations pertaining to the design and construction of dams because each dam is unique and as such must be designed using current good engineering design practices. Each design is submitted to and reviewed by the State Engineer's staff prior to acceptance by the State Engineer.

Following the State Engineer's endorsement of approval on the application and prior to commencement of construction, the owner nominates an engineer registered in New Mexico to supervise construction of the dam. The State Engineer reviews the qualifications of the engineer and if acceptable he issues a letter approving the engineer and setting forth conditions under which he will supervise construction.

For the design of the spillway, we require that it be sized in accordance with the minimum emergency spillway hydrologic criteria set forth in U.S. Department of Agriculture Technical Release No. 60, June 1976 (revised August 1981).

The State Engineer has taken the position that criteria now deemed appropriate for new structures may not be appropriate for existing structures. As a practical matter it may be necessary to entertain a greater risk with existing structures than would be acceptable in the design of new structures. We have undertaken to consult with engineers in federal and other state governments and in private practice to assess what best represents good engineering judgment. The staff of the State Engineer has undertaken breach analysis studies for selected high hazard dams in New Mexico. The results of these studies are used to better assess the requirements for emergency spillway design of existing and new dams in New Mexico.

New York
(From letter dated May 30, 1984)

The following is extracted from material supplied by the Chief, Dam Safety Section, New York State Department of Environmental Conservation, who has described the state's spillway criteria as follows:

Dams in New York are classified according to downstream hazard classification. Class "A" is the lowest hazard class and class "C" is the highest. Failure of a class "C" dam could result in loss of life. With regard to hydrologic design criteria we follow the following standards:

New Earth Dams
  Class "A"
  1. Small Dam
  Spillway shall have sufficient capacity to discharge a Spillway Design

Flood equal to a 100 year storm and also maintain a minimum freeboard of one foot between design high water and the top of dam.

2. Large Dam

Spillway shall have sufficient capacity to discharge a Spillway Design Flood equal to 150% of the 100 year storm and also maintain two feet of freeboard between design high water and the top of dam.

Class "B"

1. Small Dam

Spillway shall have sufficient capacity to discharge a Spillway Design Flood equal to 225% of the 100 year storm and also maintain a minimum freeboard of one foot.

2. Large Dam

Spillway shall have sufficient capacity to discharge a Spillway Design Flood equal to 40% of the probable maximum flood and also maintain two feet of freeboard.

Class "C"

1. Small Dam

Spillway shall have sufficient capacity to discharge a Spillway Design Flood equal to one-half of the probable maximum flood and also maintain one foot of freeboard.

2. Large Dam

Spillway shall have sufficient capacity to discharge a Spillway Design Flood equal to the probable maximum flood and also maintain two feet of freeboard.

Small Dam Definition

a. Height of Dam equal to or less than 40 feet

b. Storage at normal water surface equal to or less than 1000 acre feet

Large Dam Definition

a. Height of Dam greater than 40 feet

b. Storage at normal water surface equal to or less than 1000 acre feet

New Concrete or Masonry Dams

Dams will be designed for the same Spillway Design Floods as indicated for new earth dams. However, for concrete or masonry gravity dams overtopping will be acceptable provided that the spillway and nonoverflow section will be able to meet the structural stability requirements with regard to sliding and overturning.

Existing Dams

1. *Class "A"*

Shall have adequate capacity to discharge a Spillway Design Flood equal to a 100 year storm.

2. *Class "B"*
Shall have adequate capacity to discharge a Spillway Design Flood equal to 150% of the 100 year storm.

3. *Class "C"*
Shall have adequate capacity to discharge a Spillway Design Flood equal to one-half of the probable maximum flood.

North Carolina
(From letter dated May 18, 1984)

The State's Dam Safety Regulations include Tables A-9 and A-10. In addition, two other criteria for sizing spillways are set out as follows:

(1) Within 15 days following passage of the design storm peak the spillway system shall be capable of removing from the reservoir at least 80% of the water temporarily detained in the reservoir above the elevation of the primary spillways.

(2) Rational selection of a safe spillway design flood for specific site conditions based on a quantitative analysis is acceptable. The spillway should be sized so that the increased downstream damage resulting from overtopping failure of the dam would not be significant as compared with damage by the same flood in the absence of overtopping failure. A design storm more frequent than once in 100 years will not be acceptable for any class C dam. (The state normally requires that the assumed time of breach development in such analysis be in the range of 15 to 30 minutes.)

TABLE A-9 Criteria for Spillway Design Storms[a]

Size	Size Classification	
	Total Storage (ac-ft)[a]	Height (ft)[a]
Small	Less than 750	Less than 35
Medium	Equal to or greater than 750 and less than 7,500	Equal to or greater than 35 and less than 50
Large	Equal to or greater than 7,500 and less than 50,000	Equal to or greater than 50 and less than 100
Very large	Equal to or greater than 50,000	Equal to or greater than 100

[a]The factor determining the largest size shall govern.

TABLE A-10   Minimum Spillway Design Storms

Hazard	Size	Spillway Design Flood (SDF)
Low (Class A)	Small	50 yr
	Medium	100 yr
	Large	1/3 PMP
	Very large	1/2 PMP
Intermediate (Class B)	Small	100 yr
	Medium	1/3 PMP
	Large	1/2 PMP
	Very large	3/4 PMP
High (Class C)	Small	1/3 PMP
	Medium	1/2 PMP
	Large	3/4 PMP
	Very large	PMP

North Dakota
(From letter dated May 29, 1984)

A state official reports that North Dakota is in process of developing safety criteria for dam design. It is anticipated that the criteria will give consideration to the size of dams, downstream hazard categories, probable effects of dam failures, and other pertinent national or man-made conditions. Also, it is anticipated that the hazard classifications and hydraulic analysis guidelines will follow the criteria established for the U.S. Army Corps of Engineers National Dam Inspection Program. Proposed classifications of dams are as follows:

Category	Storage Capacity (ac-ft)	Height (ft)
1. Large	50,000 ac-ft and larger	100 and higher
2. Intermediate	1,000 through 49,999	40 through 99
3. Small	50 through 999	25 through 39
4. Very Small	12½ through 49	8 through 24
5. Pond	½ through 12.4	2 through 7

Ohio
(From letter dated June 5, 1984)

The Ohio Administrative Rules relating to dam safety provide for four classes of dams and corresponding spillway design criteria as follows:

1. When failure of the dam would result in probable loss of human life or serious hazard to health, serious damage to homes, high-value industrial or

commercial properties, or major public utilities, the dam shall be placed in class I. Dams having a storage volume greater than five thousand acre-feet or a height of greater than sixty feet shall be placed in class I.

2. When failure of the dam would result in a possible health hazard or probable loss of high-value property or damage to major highways, railroads, or other public utilities, but loss of human life is not envisioned, the dam shall be placed in class II. Dams having a storage volume greater than five hundred acre-feet or a height of greater than forty feet shall be placed in class II.

3. When failure of the dam would result in property losses restricted mainly to rural lands and buildings and local roads, and no loss of human life or hazard to health is envisioned, the dam shall be placed in class III. Dams having a height of greater than twenty-five feet, or a storage volume of greater than fifty acre-feet, shall be placed in class III.

4. When failure of the dam would result in property losses restricted mainly to the dam and rural lands, and no loss of human life or hazard to health is envisioned, the dam may be placed in class IV. Dams which are twenty-five feet or less in height and have a storage volume of fifty acre-feet or less, or dams, regardless of height, which have a storage volume of fifteen acre-feet or less, may be placed in class IV. No proposed dam shall be placed in class IV unless the applicant has submitted the preliminary design report required by rule 1501:21-5-02 of the Administrative Code.

The magnitude of the design flood shall be determined from actual streamflow and flood frequency records or from synthetic hydrologic criteria based on current publications prepared by the Ohio Division of Water, the United States Army Corps of Engineers, the United States Geological Survey, the National Oceanographic and Atmospheric Administration, or others acceptable to the Chief of the Division of Water. The minimum design flood will be:

1. For class I dams, the probable maximum flood;
2. For class II dams, fifty per cent of the probable maximum flood; and
3. For class III dams, twenty-five per cent of the probable maximum flood.

(The Administrative Rules give no minimum design flood for Class IV dams.)

Pennsylvania
(From letter dated June 5, 1984)

Hydraulic criteria for dam safety are shown by the excerpts reproduced by the following extracts (Tables A-11 and A-12) from Pennsylvania's Rules and Regulations.

TABLE A-11   Size Classification (Pennsylvania)

Class	Impoundment Storage (ac-ft)	Dam Height (ft)
A	Equal to or greater than 50,000	Equal to or greater than 100
B	Less than 50,000 but greater than 1,000	Less than 100 but greater than 40
C	Equal to or less than 1,000	Equal to or less than 40

NOTE: Size classification may be determined by either storage or height of structure, whichever gives the higher category.

TABLE A-12   Hazard Potential Classification (Pennsylvania)

Category	Loss of Life	Economic Loss
1	Substantial	Excessive (extensive residential, commercial, agricultural, and substantial public inconvenience)
2	Few (no rural communities or urban developments and no more than a small number of habitable structures)	Appreciable (damage to private or public property and short duration public inconvenience)
3	None expected (no permanent structure for human habitation)	Minimal (undeveloped or occasional structures with no significant effect on public convenience)

The design flood criteria set out in Pennsylvania's regulations are as follows:

Size and Hazard Potential Classification	Design Flood
A-1, A-2, B-1	PMF
A-3, B-2, C-1	½ PMF to PMF
B-3, C-2	100 year to ½ PMF
C-3	50 year to 100 year freq.

The Department may, in its discretion, require consideration of a minimum design flood for any class of dams or reservoirs in excess of that set forth above where it can be demonstrated that such a design flood requirement is necessary and appropriate to provide for the integrity of the dam or reservoir and to protect life and property with an adequate margin of safety.

The Department may, in its discretion, consider a reduced design flood for any class of dams or reservoirs where it can be demonstrated that such design flood provides for the integrity of the dam or reservoir and protects life and property with an adequate margin of safety. The regulations provide, also, that the Probable Maximum Flood (PMF) is to be derived from Probable Maximum Precipitation (PMP) estimates obtained from the National Weather Service of the National Oceanographic and Atmospheric Administration (NOAA).

The Chief, Division of Dam Safety in Pennsylvania, reports that changes in technique in application of PMP estimates advocated in recent NOAA reports have caused problems with dam owners.

South Carolina
(From letter dated May 17, 1984)

Requirements for spillway capacities in the South Carolina dam safety regulations are as shown in Table A-13.

South Dakota
(From letter dated May 16, 1984)

State does not have a dam safety program but did inspect high hazard dams under the Corps of Engineers national dam inspection program. State has no dam safety criteria.

Texas
(From letter dated May 25, 1984)

Published rules of the Texas Water Development Board have not specified hydraulic criteria for dams. Criteria of such authorities as SCS and U.S. Army Corps of Engineers have been used. However, changes in the rules to include hydraulic criteria for dams are being developed.

Utah
(From letter dated May 8, 1984)

The following material on spillway hydrology is quoted from "Rules and Regulations Governing Dam Safety in Utah," dated January 1982:

TABLE A-13   Spillway Design Flood Criteria

Hazard	Size	Spillway Design Flood (SDF)
High	Very small	100 yr to 1/2 PMF
	Small	1/2 PMF to PMF
	Intermediate	PMF
	Large	PMF
Significant	Very small	50 to 100 yr frequency
	Small	100 yr to 1/2 PMF
	Intermediate	1/2 PMF to PMF
	Large	PMF
Low	Small	50 to 100 yr frequency
	Intermediate	100 yr to 1/2 PMF
	Large	1/2 PMF to PMF

NOTE: When appropriate, the spillway design flood may be reduced to the spillway discharge at which dam failure will not significantly increase the downstream hazard which exists just prior to dam failure.

Unless specifically exempted by the State Engineer, the spillway design calculations shall follow the list below. The spillway shall be sized such that the appropriate flood can pass through the structure without overtopping.

High hazard	1/2 PMF — PMF
Moderate hazard	100-year frequency — 1/2 PMF
Low hazard	100-year frequency

Virginia
(From letter dated June 14, 1984)

The following material (Table A-14) is extracted from material furnished by the Dam Safety Section, State Water Control Board:

Table A-14 defines the appropriate spillway design flood. This is essentially the same as the Guidelines of the Corps of Engineers. Presently an amendment is proposed to permit engineering judgment on the appropriate spillway design flood since the PMP is constantly changing.

There are many factors involved in the Spillway Design Flood in addition to the capacity and height. The watershed area and the slope should also be considered. It may be well to consider more steps as 1/4 PMF, 1/2 PMF and 3/4 PMF.

TABLE A-14

Class of Dam	Hazard Potential If Impounding Structure Fails	Size Classification			Spillway Design Flood (SDF)[b]
		Maximum Capacity (ac-ft)[a]	Height	(ft)[a]	
I	Probable loss of life; excessive economic loss	Large	>50,000	>100	PMF
		Medium	>1,000 and <50,000	>40 and <100	PMF
		Small	>50 and <1,000	>25 and <40	1/2 PMF to PMF
II	Probable loss of life; appreciable economic loss	Large	>50,000	>100	PMF
		Medium	>1,000 and <50,000	>40 and <100	1/2 PMF to PMF
		Small	>50 and <1,000	>25 and <40	100-yr to 1/2 PMF
III	No loss of life expected; minimal economic loss[c]	Large	>50,000	>100	PMF
		Medium	>1,000 and <50,000	>40 and <100	1/2 PMF to PMF
		Small	>50 and <1,000	>25 and <40	50 yr to 100 yr

[a]The factor determining the largest size shall govern.

[b]The recommended design floods in this column represent the magnitude of the spillway design flood (SDF), which is intended to represent the largest flood that need be considered in the evaluation of a given project, regardless of whether a spillway is provided; i.e., a given project should be capable of safely passing the appropriate SDF. Where a range of SDF is indicated, the magnitude that most closely relates to the involved risk should be selected.

[c]Class III impounding structures for agricultural purposes less than or equal to 100 acre-feet in capacity and less than or equal to 25 feet in height are exempt from regulation per Section 2.01-d-iii upon certification by the owner per Section 2.01q.

PMF: Probable Maximum Flood. The flood that may be expected from the most severe combination of critical meteorologic and hydrologic conditions that are reasonably possible in the region. The PMF is derived from the current probable maximum precipitation (PMP), which information is generally available from the National Weather Service, NOAA. Most federal agencies apply reduction factors to the PMP when appropriate. Reductions may be applied because rainfall isohyetals are unlikely to conform to the exact shape of the drainage basin. In some cases local topography will cause changes from the generalized PMP values; therefore, it may be advisable to contact federal construction agencies to obtain the prevailing practice in specific cases.

100-Year: 100-Year Exceedance Interval. The flood magnitude expected to be exceeded, on the average, once in 100 years. It may also be expressed as an exceedance frequency with a one percent chance of being exceeded in any given year.

Washington
(From letter dated June 6, 1984)

The Supervisor, Dam Safety Section, reports that his section has been engaged for the past three years in regional frequency analysis of precipitation data for the purpose of selecting design storms to be used in the computation of inflow design floods for spillway design. The decision to use this probabilistic approach was based on the perception that basing designs on PMP or percentages of PMP leads to drastically different levels of safety throughout Washington State.

It is envisioned that the adopted spillway design guidelines will include two components: (1) a downstream hazard assessment weighting, and (2) a regional precipitation analysis to provide probabilistic information on the magnitude, frequency and temporal distribution of rainfalls within extreme events.

West Virginia
(From letter dated May 29, 1984)

The following are extracts from "Dam Control Regulations," effective February 1, 1982, issued by the West Virginia Department of Natural Resources:

Hazard Classification

The hazard potential shall be determined by the applicant based on the potential loss that would result due to a failure and the classification determined as listed below:

(a) Class A—Dams located in rural or agricultural areas where failure may damage farm buildings, agricultural land, or secondary highways. Failure of the structure would cause only loss of the structure and loss of property use such as related roads, but with little additional damage to adjacent property. Any impoundment exceeding 25 feet in height or 200 acre-feet storage volume or having a watershed exceeding 500 acres shall not be a Class A structure.

(b) Class B—Dams located in predominantly rural agricultural areas where failure may damage isolated homes, primary highways or minor railroads or cause interruption of relatively important public utilities. Failure of the structure may cause great damage to property and project operations.

(c) Class C—Dams located where failure may cause loss of human life, serious damage to homes, industrial and commercial buildings, important public utilities, primary highways, or main railroads. This classification must be used if failure would cause possible loss of human life.

Design Requirements

Design Storm—All dams shall be designed to meet the following minimum hydrologic criteria based on hazard classification:

(1) Class A dams shall be designed for a minimum of $P_{100} + 0.12(PMP-P_{100})$ inches of rainfall in six (6) hours plus three (3) feet of freeboard. If the storage × effective height is less than 3,000 (acre-feet × feet) then Soil Conservation Pond Standard 378 may be substituted.

(2) Class B dams shall be designed for a minimum of $P_{100} + 0.40(PMP-P_{100})$ inches of rainfall in six (6) hours plus three (3) feet of freeboard.

(3) Class C dams shall be designed for the probable maximum precipitation, or for 80 percent of the probable maximum precipitation plus three (3) feet of freeboard provided the watershed is less than ten (10) square miles in area.

PART 3—OTHER GOVERNMENTAL AGENCIES

City of Los Angeles, California, Department of Water and Power
(From letter dated July 3, 1984)

The following is adapted from a list of design procedures and criteria relating to extreme floods furnished by the Department of Water and Power:

1. Develop the PMP storm using the procedures in "Hydrometeorological Report No. 36—Interim Report—Probable Maximum Precipitation in California" (U.S. Weather Bureau, 1961a) or "Hydrometeorological Report No. 49—Probable Maximum Precipitation Estimates, Colorado River and Great Basin Drainages" (Hansen et al., 1977).

2. Calculate runoff generated by the Probable Maximum Precipitation (PMP) storm using the latest version of the HEC-1 Flood Hydrograph Package computer program developed by the U.S. Army Corps of Engineers.

3. Provide adequate spillway capacity to accommodate the PMP storm.

4. Provide adequate freeboard to accommodate the runoff generated by the PMP storm.

5. Provide sufficient storm water bypass facilities for off-stream reservoirs.

6. Provide sufficient blow off capability.

7. Meet the requirements of the California Department of Water Resources, Division of Safety of Dams.

East Bay Municipal Utility District, California
(From letter dated May 25, 1984)

The East Bay Municipal Utility District (EBMUD) owns a number of dams. EBMUD's manager reports that the District does not have formalized or written criteria for dams but has attempted to apply "state-of-the-art" criteria, standards and procedures in both the design of new facilities and the analysis of existing facilities.

Salt River Project, Arizona
(From letter dated June 25, 1984)

The following two paragraphs are quoted from letter of the General Manager of the Salt River Project (SRP):

SRP operates and maintains six (6) large high-hazard dams upstream of metropolitan Phoenix. Current studies by the USBR and Corps of Engineers have revised both the hydrologic and seismic design criteria for these structures causing them all to be categorized as unsafe due to their inability to safely accommodate the new Inflow Design Floods (IDF) and Maximum Credible Earthquakes (MCE). Recently completed studies also indicate that failure of these dams could result in the inundation of as many as a quarter million (250,000) Phoenix Valley residents. Although legal title and ultimate Safety of Dams responsibility rests with the USBR, SRP senses a strong obligation to investigate and support all efforts to insure the safety of the dams on the Salt and Verde rivers.

Since the Salt River Project does not own the Salt and Verde River dams, it does not establish the hydrologic and seismic criteria but rather operates under the criteria set by the dam owner, i.e. the USBR. SRP has, however, conducted several studies in the past which are directly or closely related to the concerns being investigated by the National Research Council study.

Information on the following described studies was furnished by the General Manager:

1. *Synopsis of Selected USBR Inflow Design Floods for Large Dams*, PRC Engineering Consultants, Inc., May 1980. The study was undertaken to compile a synopsis of U.S. Bureau of Reclamation (USBR) hydrologic design criteria to assist the Salt River Project (SRP) in assessing the hydrologic

degree of risk associated with the six storage dams that comprise the SRP system.

2. *Paleoflood Hydrology Studies on Salt and Verde Rivers,* Dr. Victor Baker, University of Arizona. This study is currently in progress and is intended to provide additional information on the magnitude and frequency of historic and prehistoric floods on the Salt and Verde rivers. A letter report identifies the cursory results obtained during a reconnaissance investigation and a glimpse of the information hoped to be obtained during the full scale study. This study is scheduled to be completed by October 1, 1984.

The report on the first-listed study traces the hydrologic design practices of the U.S. Bureau of Reclamation from the establishment of the Bureau in 1902 till the present and experience of the Bureau in 7,946 dam-years of accumulated exposure at 259 storage dams. The Gibson Dam on the Sun River in Montana was the only Bureau dam overtopped in that period of record. The report noted that estimates of reservoir inflow at the Roosevelt and Horseshoe dams of the Salt River Project for currently-used design rainfalls greatly exceed the spillway capacities available at those dams.

The letter relating to the second-listed study stated that brief reconnaissance of pre-historic flood deposits along the Salt and Verde rivers had revealed radiocarbon-datable materials. It concluded that an extensive paleoflood record existed that probably could be used to evaluate the sequence of largest floods over several thousand years.

Santee Cooper
(South Carolina Public Service Authority)
(From letter dated July 12, 1984)

The following is quoted from letter of the President, Santee Cooper:

We are very interested in the establishment of acceptable risk levels for seismicity as well as for spillway adequacy in Federal dams.

As for our design procedures, criteria and standards for dam safety and inspections, Santee Cooper is licensed by the Federal Energy Regulatory Commission (FERC) and follows FERC guidelines regarding dam safety.

PART 4—TECHNICAL SOCIETIES

American Society of Civil Engineers (ASCE)

On May 9, 1981, the ASCE Board of Directors adopted a policy statement entitled "Responsibility for Dam Safety." As the title indicates, the statement was directed primarily at the placement of responsibility for the safety of dams but it did contain references to design criteria. The following are extracts from that statement:

In the development of a project involving a dam, there are uncertainties in predicting the natural events to which the dam will be subjected, in estimating future project effects and benefits, and in forecasting the performance of project components. Available design alternatives may involve varying first costs and degrees of risk to the owner. Where only costs to the owner are involved, economic analyses based on historical records of natural events and performance records of similar components at other projects provide bases for selection among design alternatives. However, if possible loss of human life, loss of vital community services or extensive damages to others may be involved, the adopted design should seek to minimize the potential for such losses or damages.

Hydrologic relationships used in operations studies to evaluate project effects and to design project components should be based on thorough analyses of local and regional hydrologic records. The theoretical probable maximum flood (PMF) potential of the basin should be considered in designing any dam where there would be significant hazard to lives and property in downstream areas.

International Commission on Large Dams (ICOLD)

The following is extracted from a draft dated February 1984 of a proposed "Guidelines on Dam Safety" prepared by the ICOLD Committee on Dam Safety:

As a general rule, the design of the dam and reservoir shall be based on the probable maximum flood. The latter shall derive from the combination of maximum runoff volumes with most unfavorable runoff conditions and is to be used to produce the design flood hydrograph. The capacity of gated spillways shall be sufficient to discharge the full design flood without taking into account the dampening effect resulting from flood routing through the reservoir. A reduction of the design flood as derived from the probable maximum flood, or the consideration of the effect of flood routing when determining the spillway capacity, should be permitted under especially favorable conditions. Such conditions may be:

—The permanent availability of reserve storage capacity of the reservoir, between the normal top water level and the maximum reservoir level, compatible with the temporary surcharge volume deriving from the partial retention of the inflowing flood. The availability of the mentioned reserve storage capacity must be combined with highly reliable operating procedures that assure the opening of the spillway gates in accordance with the predetermined flood routing program.

—The existence of an additional fuse plug type spillway the rupture of which would not increase the downstream flood beyond the acceptable risks.

—A permanently warranted low downstream risk level that should at no time include any risk to human life.

—Other favorable circumstances that permit the exemption from the above mentioned requirements, in accordance with criteria and regulations established by the Government Agency.

In any case, however, the determination of design flood and spillway capacity of all dams within the same drainage area must be based on uniform criteria and procedures.

U.S. Committee on Large Dams (USCOLD)

This United States component of ICOLD has not attempted to promulgate design standards for dams, but in the late 1960s an USCOLD working committee undertook a survey of design practices in the United States for sizing spillways. The results of that survey, which was accomplished by use of a questionnaire, were presented in a 1970 USCOLD report "Criteria and Practices Utilized in Determining the Required Capacity of Spillways." The following material is extracted from that report:

All respondents stated that current policies of their agencies or firms were consistent with the following general statement:

"When a high dam, capable of impounding large quantities of water, is constructed above a populated community, a distinct hazard to that community from possible failure of the dam is created unless due care is exercised in every phase of the engineering design, construction, and operation of the project to assure complete safety. The policy of deliberately accepting a recognizable major risk in the design of a high dam simply to reduce the cost of the structure has been generally discredited from the ethical and public welfare standpoint, if the results of a failure would imperil the lives and lifesavings of the populace of the downstream flood plain. Legal and financial capabilities to compensate for economic losses associated with major dam failures are generally considered as inadequate justifications for accepting such risks, particularly when severe hazards to life are involved. Accordingly, it is the policy of this agency that high dams impounding large volumes of water be designed to conform with Security Standard 1."

1. *Standard 1*

a. High dam impounding large volumes of water, sudden release of which would create major hazards to life or property downstream.

b. Dams of such economic importance that prevention of overtopping during extreme floods including the probable maximum flood is of such importance as to justify the expenditures required, notwithstanding the low probability of occurrence of overtopping.

2. *Standard 2*

It is recognized that some low head dams can be overtopped without failing or if they fail a hazardous flood wave will not result downstream. In such cases the design capacity of the spillway and related features may be based largely on economic consideration. Typical applications include:

a. Dams specifically designed so that overtopping will not cause either failure or serious damage downstream.

b. Run-of-river hydroelectric power or navigation dams, diversion dams, and similar structures where relatively small difference between headwater and tailwater elevations will prevail during overtopping floods and where the cost of preventing overtopping is high in comparison with economic losses otherwise probable.

c. Sub-impoundment dams adjacent to larger reservoirs, where possible release from breaching can be absorbed by the larger reservoir without major hazard.

3. *Standard 3*

a. Dams impounding a few thousand acre-feet or less, so designed as to assure a relatively slow rate of failure if overtopped and located where hazard to life and property in the event of dam failure would clearly be within acceptable limits. Under certain circumstances, with special precautions larger dams may fall under this standard.

b. Sub-impoundment dams of the nature described under Standard 2c.

4. *Standard 4*

a. Dams forming small recreational lakes or water supply ponds located where the probability of serious property damage would be acceptably small.

b. Dams forming relatively small farm ponds where failure would not constitute a serious hazard downstream.

TABLE A-15   Number of Projects Designed or Constructed by Reporting Agencies (Approximate Estimates)—From 1970 USCOLD Report

Reporting Agency or Firm	Standard 1	Standard 2	Standard 3	Total
Federal agencies				
Corps of Engineers, USA	320	70	10	400
Bureau of Reclamation	195	134	0	329
Tennessee Valley Authority	22	8	8	38
Soil Conservation Service	560	—	4,780	5,340
Subtotals (Federal)	1,097	212	4,798	6,107
State agencies and private engineering firms	295	64	18	377
Totals	1,392	276	4,816	6,484

The primary objective of the survey was to compile information concerning practices and criteria actually used in the design of existing dams and those scheduled for construction in the near future. Accordingly, each recipient of the questionnaire was requested to indicate the number of projects covered in his reply, identified according to Security Standards 1, 2 and 3 defined above in order that the scope of application of various procedures and criteria might be evaluated. Table A-15 is a breakdown of projects reported in the various classifications.

PART 5—FIRMS IN UNITED STATES

Acres American, Inc., Buffalo, New York
(From letter dated July 7, 1984)

The following extract from a paper prepared for a seminar summarizes the practices of the Acres American organization in determining spillway capacities:

Like most organizations, Acres has not adopted a rigid position on spillway capacity criteria—circumstances alter cases.

General statement for large reservoirs is:

—Check carefully for largest flood types (spring snowmelt, hurricane, other storm rainfall);

—Design spillway to pass 10,000-year flood with no reservoir surcharge, all gates in operation, no power turbines in use;

—Route flood through drawn down reservoir, if drawdown will *always* be accomplished by time of flood (e.g. snowmelt flood);

—Verify that MPF (Maximum Probable Flood) can be handled without major damage or loss of life, through the use of freeboard for storage and/or fuse plug spillways, or other emergency spillways.

Alabama Power Co., Birmingham, Alabama
(From letter dated July 18, 1984)

Alabama Power Company supplied information on hydrologic studies now under way of eleven projects in the Coosa and Tallapoosa river basins. In the PMF determinations the company is transposing two actual storm rainfall patterns, the Yankeetown, Florida, storm of September 1950 and the Elba, Alabama, storm of March 1929, adjusted in accord with Hydromet practice, in lieu of using PMP estimates from the U.S. Weather Service. It is the company's position that such use of transposed and adjusted rainfalls will come closer to depicting actual conditions to be expected in the basin during such intense storms. Company's projects must meet FERC standards.

R.W. Beck and Associates, Seattle, Washington
(From letter dated June 12, 1984)

The following is quoted from the firm's letter:

Beck generally has followed the U.S. Army Corps of Engineers (COE) criteria for severe hydrologic events by developing the Probable Maximum Flood (PMF) from the Probable Maximum Precipitation (PMP) and applying COE hazard criteria to select the Spillway Design Flood (SDF). Most State and Federal agencies have accepted the Corps approach as being conservative, and only in special circumstances involving unimportant structures where substantial savings can be realized in analysis and engineering are simplified methodologies employed by Beck.

Central Maine Power Company, Augusta, Maine
(From letter dated July 31, 1984)

The Central Maine Power Company has supplied data sheets pertaining to structural analyses for five of its hydroelectric power projects. The analyses were made by Charles T. Main, Inc. The data sheets are not explicit in regard to hydrologic criteria used but do indicate that a "probable maximum flood" was used in the structural analyses. Company's projects are subject to FERC regulations.

Duke Power Company, Charlotte, North Carolina
(From letter dated July 19, 1984)

Information supplied by Duke Power Company indicates that its standards for dams are comprised of the regulations of the Federal Energy Regulatory Commission supplemented by standards and criteria issued by a number of federal and state agencies.

Charles T. Main, Inc., Boston, Massachusetts
(From information furnished by Llewellyn L. Cross, June 18, 1984)

In serving Main's various clients, who are scattered about the world, all of the standard hydrologic techniques are employed.

In the U.S. and other areas where the Probable Maximum Flood is mandated as the design standard, the applicable Hydrometeorological Reports are used. Where these are not available, a hydrometeorological approach using precipitable water and dew points is taken. Storm transposition and maximization techniques are also employed.

Unit hydrographs are derived from historically appropriate flood events where the data are available. In cases of no records, unit hydrographs are developed from the physical characteristics of the basin.

Diversion floods are computed using statistical methods adapted to site-specific situations.

In many instances, for projects in remote areas having no data, storm models appropriate to the catchment are developed using meteorological methods and parameters. These models are then maximized for rainfall intensity and duration and critically sited on the project catchment.

For many cases, the spillway design flood has been the result of snow melt and this has resulted in the development of necessarily crude models relating snow melt to incremental melt temperature.

Planning Research Corporation (PRC), Denver, Colorado
(From letter dated June 19, 1984)

The following is extracted from a description of the hydrologic criteria used by PRC:

We normally follow the generally accepted design criteria that, if the failure of a water storage dam *could* result in loss of life or substantial loss of property, the dam and spillway should be sized to safely pass the Probable Maximum Flood (PMF). For projects where loss of life or substantial property loss will not be a consequence of a dam failure, then a lesser flood is used as the Inflow Design Flood (IDF). The size of the IDF is site specific for each project, but we never use anything less than the 100-year event.

In the United States, the magnitude of the project IDF is almost always set by regulation (State Engineers Office or some other State or Federal Agency). Overseas, however, the decision with regard to the magnitude of the IDF is the responsibility of the engineer. We always present our recommendation to our client, discuss it with him and reach agreement at an early stage of the project.

The majority of our projects include major dams to supply water to large irrigation or hydropower developments and, therefore, we normally use the PMF as the Inflow Design Flood.

At times, we believe it is in the public's best interest to take a different approach to establishing the project inflow design flood. In some instances, the routed PMF outflows from the project spillway are so great that significant damage will take place as a result of those outflows even without the occurrence of a dam failure. Also, if one considers the incremental downstream flood hazard resulting from a dam break, compared to an existing condition during the same flood event, the additional flooding, and therefore flood damage, may prove to be insignificant. If a review of the proposed project features and downstream topographic conditions indicates that a dam failure would result in insignificant incremental damages, then we might propose that a dam break analysis be performed, and that consideration be given to designing for an IDF which is smaller than the PMF, thus attempting to optimize project cost and risk. One must use caution in considering the use of this approach, however, because the results of a dam break analysis are highly dependent on assumptions made concerning the time of failure, the mode of failure and the downstream topographic conditions. For example, I know of an instance where a 25-foot high dam resulted in a 70-foot high downstream flood wave. This occurred because the valley downstream was relatively narrow and heavily wooded, resulting in debris dams being formed downstream during the flooding, and those dams resulted in temporary ponding and then failed suddenly.

Yankee Atomic Electric Co., Framingham, Massachusetts

A company representative has made available a report dated April 1984, titled "Probability of Extreme Rainfalls and the Effect on the Harriman Dam" and an early draft of the same report, dated March 1984, titled "Probability of Failure of Harriman Dam due to Overtopping." These reports describe studies of a 60-year-old hydroelectric power project in Vermont in the upper Deerfield River basin, which is upstream of the site of the Yankee atomic power development. As part of the study of safety of the atomic power installation, the Nuclear Regulatory Commission has required an assessment of the failure potential of the upstream dam.

The studies of the flood-producing potentials of the 200-square-mile drainage area of Harriman Dam had three aspects of considerable pertinence to the present effort of the Committee on Criteria for Dam Safety: (1) the range in the estimates for probable maximum precipitation (PMP) over the area, (2) the use of what was termed the "unconditional probability approach" in developing estimates of average frequency of return for extremely large rainfalls, and (3) the development of estimates of probability of dam failure by overtopping with various confidence levels. The 24-hour, 200-square-mile PMP estimates ranged from 14.3 inches to over 22 inches.

The "unconditional probability approach" is described in the following quotation from the April 1984 report:

"In the unconditional probability approach, no *a priori* assumption was made concerning the mathematical form of the statistical distribution. In its simplest sense, the probability of exceeding a particular rainfall depth at a point of interest is estimated by multiplying the annual frequency of the events of such depth occurring anywhere within a large zone of interest times the probability that that event will occur directly over a specific point of interest. The former annual frequency can be calculated from the historical records. The latter probability of the event occurring over a specific location can be estimated simply as the ratio of the average storm area in which a depth is equaled or exceeded to the total area of the large zone of interest."

In applying this approach, the annual frequencies of 24-hour rainfalls equaling or exceeding various depths above 6 inches over any 200-square-mile area within each of a number of geographical zones were developed from historical records. A total of seven zones were used (ranging in total area from 36,783 square miles to 249,372 square miles), and each zone contained the 200-square-mile area upstream from Harriman Dam. The frequencies for occurrence over any 200-square mile area within each geographical zone were converted to estimated probabilities for occurrence over the drainage area above Harriman Dam by simple ratios of the target areas involved. Thus a rainfall with annual frequency of 0.01 over any 200-square-mile area within the largest 249,372-square-mile zone would have an estimated annual probability of occurrence over the drainage area of Harriman Dam of $0.01 \times 200/249,372 = 0.000008$, or, to put this in terms in common use, the 100-year rainfall for any 200-square-mile area in the zone becomes the 125,000-year rainfall for the area upstream from the dam. This conversion is based on these assumptions:

1. The approximately 100-year period in New England for which results of depth-area-duration studies for all major storms are available is representative of long time averages.
2. The geographic zones used are meteorologically homogeneous.
3. Occurrence of a major rainfall over a specific target area is a random chance event.

By the "unconditional probability approach," the annual probabilities of the PMP estimates for the drainage area of Harriman Dam were assessed as follows:

24-hour PMP	Annual Probability
14.3″	3.5×10^{-5}
22 + ″	2.2×10^{-7}

The Yankee Atomic Electric Company's report states that the Nuclear Regulatory Commission generally has accepted, as a basis for design, seismic hazard curves with annual probabilities of 10^{-3} to 10^{-4} and implies that hydrologic design events with similar probabilities should be reasonable bases for design.

PART 6—OTHER ENTITIES IN UNITED STATES

Illinois Association of Lake Communities
(From letter dated July 19, 1984)

The President, Illinois Association of Lake Communities, stated that he was writing on behalf of the communities of the association and other municipal dam operators within the state whose dams have been inspected under the National Dam Inspection Program of the U.S. Army Corps of Engineers and found to have inadequate spillway capacity under the criteria used for that program. He protested any requirement that operators of dams, for which construction permits were originally issued and which are being operated and maintained in a safe, reliable manner, be required to meet new dam safety criteria. He emphasized the costs of upgrading such dams, stated such costs could mean potential bankruptcy for home owner associations, and suggested it would be senseless and unrealistic to require spillway designs for 26″ of rain in a six-hour period.

A separate communication of same date from a law firm representing the

Association (McDermott, Will & Emory) questions the legality of requiring application of PMF flood criteria to existing dams. The following bases of argument were presented.

a. Retroactive application of PMF criteria for existing dams would be a violation of the constitutional rights of the dam owners.

b. The classification of a dam as "high hazard" based only on the location of the dam is a "conclusive and irrebuttable presumption" that is violative of due process rights of the owners.

c. A system of regulation of dams not based on the actual condition of existing dams is not reasonably related to the purpose of protecting citizens from unsafe dams.

d. The application of the PMF standard to an existing dam is a taking of property without compensations.

PART 7—FOREIGN COUNTRIES

The Institution of Civil Engineers, London

In Great Britain, dam safety is entrusted to individual members of a statutory panel of engineers determined by the government to be qualified to design and inspect impoundments. After appointment as a "panel engineer," the individual may be hired by dam owners to design and inspect dams to meet statutory requirements. Each such panel engineer is personally responsible for the safety of the dams he is hired to supervise, and no mandatory standards are imposed by the government. However, to assist the panel engineers in meeting their individual responsibilities, the Institution of Civil Engineers in 1978 published a report of the Institution's Working Party on Floods and Reservoir Safety, under the title "Floods and Reservoir Safety: An Engineering Guide." Extracts from Chapter 2, "Reservoir Flood Protection Standards," of that guide follow:

Protection standards must resolve acceptably the conflicting claims of safety and economy. Although it is now considered possible to design a spillway for the total protection of a dam against overtopping, there is the clear possibility that a smaller spillway built at less expense would survive several generations without any disaster or damage occurring. However, it is not simply a matter of economic judgment . As the Institution's 1973 statement on social responsibilities states, the civil engineer should recognize the many factors which may defy expression in direct money values, particularly those which arise from effects on a community's way of life.

A crucial question when considering flood protection is the combination

TABLE A-16   Reservoir Flood and Wave Standards by Dam Category

| Category | Initial Reservoir Condition | Dam Design Flood Inflow | | | Concurrent Wind Speed and Minimum Wave Surcharge Allowance |
		General Standard	Minimum Standard If Rare Overtopping Is Tolerable	Alternative Standards If Economic Study Is Warranted	
A. Reservoirs where a breach will endanger lives in a community	Spilling long-term average daily inflow	Probable maximum flood	0.5 PMF or 10,000-yr flood (take larger)	Not applicable	Winter: maximum hourly wind once in 10 yr
B. Reservoirs where a breach (i) may endanger lives not in a community (ii) will result in extensive damage	Just full (i.e., no spill)	0.5 PMF or 10,000-yr flood (take larger)	0.3 PMF or 1,000-yr flood (take larger)	Flood with probability that minimizes spillway plus damage costs inflow not to be less than minimum standard but may exceed general standard	Summer: average annual maximum hourly wind Wave surcharge allowance not less than 0.6 m

C. Reservoirs where a breach will pose negligible risk to life and cause limited damage	Just full (i.e., no spill)	0.5 PMF or 1,000-yr flood (take larger)	0.2 PMF or 150-yr flood (take larger)		Average annual maximum hourly wind Wave surcharge allowance not less than 0.4 m
D. Special cases where no loss of life can be foreseen as a result of a breach and very limited additional flood damage will be caused	Spilling long-term average daily inflow	0.2 PMF or 150-yr flood	Not applicable	Not applicable	Average annual maximum hourly wind Wave surcharge allowance not less 0.3 m

NOTES: Where reservoir control procedure requires and discharge capacities permit, operation at or below specified levels defined throughout the year that may be adopted providing they are specified in the certificates or reports for the dam. Where a proportion of PMF is specified, it is intended that the PMF hydrograph should be computed and then all ordinates be multiplied by 0.5, 0.3, or 0.2 as indicated.

SOURCE: "Floods and Reservoir Safety: An Engineering Guide," Institution of Civil Engineers, London, 1978.

of circumstances that may arise in progressively rarer events. Three main factors have to be defined:

    (a) initial reservoir level;
    (b) flood inflow;
    (c) concurrent wind speed.

Despite continually improving techniques for defining flood hydrographs, wave run-up and flood routing, there is no indication that the engineer can do other than make separately reasoned assumptions on the levels at which the three factors listed above should be set.

In Table A-16 are set out the standards which are appropriate for the wide variety and scale of dams covered by British safety legislation. To apply them it is necessary to route the appropriate dam design flood inflow using the corresponding initial reservoir condition and to obtain two levels, one being the theoretical flood surcharge level and the other being the total surcharge level; the latter includes the appropriate allowance for wave run-up caused by the wind speed given in Table A-16 (or the minimum wave surcharge if that is greater), this wave surcharge allowance being sufficient to prevent overtopping reaching quantities that would hazard a dam crest.

Although Table A-16 may appear complex at first sight, it is designed to take account of those factors which are weighed together by panel engineers during dam inspections. Its main intentions are to ensure that, where a community could be endangered by a dam, the risk of any failure caused by a flood is virtually eliminated, but in other cases to keep expenditure to a scale justified by the risk.

Category A dams.   It is considered that public opinion will not accept conscious design for a specific threat to a community, even though it tolerates to an extent both random and accidental loss of life. Consequently, no dam above a village or town should be designed knowingly with a definite chance of a disastrous breach due to the under-provision of spillway capacity. A community defies definition in a few words; it is considered that inspection of any valley will soon reveal whether the presence of a hamlet, school or other social group means that a dam at its head should be in category A. Road and rail traffic caught in a valley flood would only accidentally be involved and would not by itself justify category A. A more difficult situation exists where an occasional camp site exists in the holiday season alongside a reservoired river; if, for example, this is in regular use by school parties it could well justify a community rating, but if it is frequented by a few unrelated short-stay individuals it need not.

Category B dams.   Category B(i) is intended to refer to inhabitants of isolated houses and, for example, to treatment plant operators in a works immediately below a dam. (These situations lend themselves to taking measures to buy out the property or to arrange flood escape routes where appropriate.) Category B(ii) refers to extensive damage, including erosion of agricultural soils and the severing of main road or rail communications.

Category C dams.   Category C covers situations with negligible risk to human life and so includes flood-threatened areas that are "inhabited" only spasmodically, e.g., footpaths across the flood plain and playing fields. In addition this category covers loss of livestock and crops.

Category D dams.   Many small reservoirs with low earth dams may cause no real problem, except that of replacement, if they wash out. These special cases, many of which are ornamental lakes kept full for aesthetic reasons, are given a separate category. A flood intense enough to cause failure of a dam would create some damage even if the valley was still in its natural state; the additional damage caused by the release of stored water may well be insignificant if the lake is small. So where the amount stored would add no more than 10% to the volume or peak of the flood it is recommended that the spillway need not pass more than the outflow from the 150 year flood (or 0.2 PMF if that is calculated more readily). The point of reference for calculating whether the dam is significant or not can be taken as the first site below the dam at which some feature of value exists (e.g., a mill or road bridge). The 1000 year flood hydrograph applicable to that catchment prior to dam construction can be used for making this 10% sensitivity test.

Economic considerations. Some reservoirs pose no threat to life but their loss would have severe economic consequences. Providing that all the losses caused by a failure can be met by remedial works and compensation payments, the sizing of the spillway and freeboard is a matter of locating the economic optimum.

Provision is made in Table A-16 for the use of an economic standard as an alternative. The strength of the least-cost method is its ability to reduce the arbitrary choice of standards which may have costly implications. However, the most economic solution over the long term may not be one that the owner can finance in the short term. Indeed the economic study itself may be expensive (although this need not always be so). The economics of the situation can be self-evident when, for example, a water treatment works is sited immediately below a dam and the loss of its output would have grave economic consequences for industrial consumers. Even for those cases where the failure of a new dam would not pose a serious threat to existing property, the additional cost of providing protection against the Probable Maximum Flood may be relatively small and it may be prudent to do so in order not to limit future development below the dam. After an economic study the panel engineer should be free to adopt safer flood control works than the nominal minimum solution if his appreciation of the extra costs of greater protection so indicates. Table A-16 contains an important qualification that the alternative economic standard should not be allowed to produce a result that involves more risk of overtopping than the minimum standard.

Design Criteria in Use for Dams Relative to Earthquake Hazards

CONTENTS

PART 3—OTHER GOVERNMENTAL AGENCIES

PART 4—PRIVATE FIRMS IN UNITED STATES

PART 1—FEDERAL AGENCIES

Ad Hoc Interagency Committee on Dam Safety of the Federal Coordinating Council for Science, Engineering, and Technology
(From "Federal Guidelines for Dam Safety," dated June 25, 1979)

The 1979 report of this group, a forerunner of the present ICODS, is directed primarily at management of organizations engaged in planning, design, construction, operation, and management of dams. The report outlines factors to be considered and procedures to be followed in investigations and design for earthquake hazards but does not specify design criteria to be followed. The report does indicate that the design seismic event is usually the maximum credible earthquake (MCE) and defines the MCE as "the hypothetical earthquake from a given source that could produce the severest vibratory ground motion at the dam."

Bureau of Reclamation, U.S. Department of the Interior
(From letter dated June 6, 1984)

The following is extracted from a description of Bureau of Reclamation's practice relating to floods and earthquakes:

Prior to the early 1970's seismic loading for Bureau dams was based on the application of 0.1g ground acceleration. In 1972, the Bureau initiated the use of the MCE (Maximum Credible Earthquake) and adopted it as the required seismic loading for Bureau dams. Under this loading, Bureau dams were required to maintain adequate stability without loss of the reservoir. The MCE associated with a specific seismic source was defined as: "the maximum earthquake that appears capable of occurring in the presently known tectonic framework. It is a rational and believable event that is in accord with all known geologic and seismologic facts. In determining the MCE, little regard is given to its *probability of occurrence*, except that its likelihood of occurring is great enough to be of concern (emphasis added)."

The Bureau currently does not have a formally adopted criteria regarding the questions of:

1. What probability of occurrence should be used in considering remote event earthquake loadings? and;
2. Under what circumstances, if any, would a seismic loading less than the hypothetical MCE's be considered for the seismic safety evaluation of a structure and how would this relate to criteria for evaluation of existing dams as opposed to new dam designs?

However, current thinking and practice with regard to these issues are described below:

In order to preserve the character of the specification of the probability as an approximate estimate; to be more conservative than the value used for housing in California; and to remain in the same realm of conservatism as the U.S. Corps of Engineers criteria, a probability of occurrence of .00002 (recurrence interval of 50,000 years) is currently considered an inclusionary criteria for remote earthquake events for Bureau seismotectonic studies. This level of improbability serves as a guide to the geologist making seismotectonic studies rather than as a strict active fault criteria.

The designation of hypothetical MCE events and their associated recurrence interval or probability of occurrence is viewed by the Bureau as a separate function from the assignment of seismic loadings for design of the structure even though often they will be one and the same. Maximum Credible Earthquakes are regarded as an actual geologic condition while (MDE) Maximum Design Earthquake (loading condition applied to the structure) is regarded as a parameter that may be specified according to the nature of the structure being considered and the consequences associated with potential seismically induced damages.

Bureau policy since 1973 has been that the MCE events are considered in the seismic evaluation of dams and that a lesser event may be used for design of noncritical structures.

In order to establish an appropriate loading level for noncritical structures on a project, as well as provide data for decision analysis studies, we are considering that seismotectonic evaluations provide a 500-year earthquake event which would be conceptually defined as the largest earthquake that is likely to occur during the life of the project and quantitatively defined as an earthquake with a recurrence interval of 500 years.

The specific seismotectonic assessment made for Bureau dam sites includes determination of:

Hypothetical Maximum Credible Earthquake(s) (MCE)—The maximum earthquake associated with relevant seismic sources is provided for each source that may produce significant earthquake shaking (greater than .05g) at the dam site. The approximate recurrence interval of each MCE is provided along with its focal depth and distance from the site. Earthquakes with a probability of occurrence up more than about .00002 are considered in establishing hypothetical MCE's.

Historic seismicity—For each source area from which an MCE is determined, an earthquake magnitude and epicentral distance is provided that represents the earthquake from that source area with a return period of 500 years if such an event is tectonically consistent with the source and able to be determined from geologic evidence, a projection of seismic evidence, or both. Otherwise such historic seismic information that is available is provided.

Surface faulting potential—The potential for surface fault rupture is assessed at each site.

Reservoir induced earthquake loading—The current procedure for evaluation of reservoir induced seismicity in the Bureau is to consider the reservoir induced event equivalent to the local MCE event. Thus, if any active or potentially active faults are located within the reservoir regime, the reservoir induced event as well as the local MCE event are defined from the capacity of that fault system. The accelerogram record would, in general, be the same whether or not reservoir induced seismicity is considered. The recognition of reservoir induced seismicity only changes the probability of occurrence or frequency of occurrence of the large magnitude event but does not change the design of the structure since the structure must be capable of handling the earthquake loading regardless of when it would occur.

The evaluation of the protection level is essential for formulating alternatives to solve the problem. This evaluation will result in one of three general cases from which to select loading conditions.

Case A—Maximum Loading Conditions

This would be the case where the level and proximity of the downstream hazard make it clear at the outset of the problem that the consequences of dam failure in terms of potential loss of life or property damage would be unacceptable regardless of how remote the chance of failure may be. Thus, the loading conditions for the various alternatives are established at the maximum level (MCE, PMF, etc.)

Case B—Loading Conditions Determined by Economic Analysis

This would be the case where the level and/or remoteness of the downstream hazard are such that it is apparent (or becomes apparent) that incremental impact of dam failure would not significantly change the potential for loss of life or other nonmonetary factors, and that an economic analysis in which the costs and benefits of reducing the hazard become the primary consideration.

Case C—Loading Conditions as a Parameter in the Ultimate Decision Making Process

This case is one where the incremental consequences of dam failure (with or without consideration of warning or other nonstructural modifications) do not clearly indicate that the dam falls under Case A or Case B. Comparison of alternatives for this case would include the economic comparison as for Case B, but would require a more comprehensive assessment of the incremental effects of dam failure on potential for loss of life (with and without warning system) as well as the incremental effects socially, environmentally, and politically for each alternative and load level.

Response Requirements for Seismic Loading

Under loading from the Maximum Design Earthquake, the structures of a project vital to the retention or release of the reservoir are required to function (1) without permitting a sudden, uncontrolled release of the reservoir and (2) without compromising the ability to make a controlled release of the reservoir.

Under loading from the 500-year earthquake (or otherwise selected Economic Design Basis Earthquake), the project facilities not critical to the retention or release of the reservoir would be designed to sustain the earthquake with repairable damage. The degree of damage which would be acceptable could be based on an economic analysis or on an estimate of the cost of the repair versus the initial cost to control the damage.

Federal Energy Regulatory Commission (FERC),
U.S. Department of Energy
(From letter dated June 12, 1984)

The following is extracted from material supplied by FERC:

The criteria presented herein apply to both the review of designs by Commission staff prior to licensing and review of licensed project by independent consultants under Part 12 of the Commission's regulations.

The review of project design for earthquake loading conditions utilizes two magnitudes of earthquakes; the maximum earthquake (ME) and the operating basis earthquake (OBE). Embankment structures should be capable of retaining the reservoir during a ME; however, deformation is acceptable. Concrete structures should be capable of performing within the elastic range during an OBE, remain operational and not require extensive repair. During a ME a concrete structure should be capable of surviving without failure of a type that would result in loss of life or excessive property damage. The earthquake design criteria is based on the Corps of Engineers ER 1110-2-1806.

Forest Service, U.S. Department of Agriculture
(From letter dated May 23, 1984)

Seismic design standards used by the Forest Service have been summarized as follows:

The Agency requires an evaluation of earth movement potential and the establishing of appropriate design criteria on a case-by-case basis. The determination of the need for detailed analyses, and subsequent design criteria, is based on factors such as the hazard presented, size of the dam and reservoir, potential ground motion at the site, site geology and the type of structure.

Interagency Committee on Dam Safety (ICODS)

An interagency task group established by ICODS has developed a draft paper entitled: "Proposed Federal Guidelines for Earthquake Analysis and Design of Dams." The parent body, ICODS, has not taken any action on the task force's draft. The draft discusses selection of design earthquake, ground motions from earthquakes, analyses of earthquake effects on dams, and evaluation of results of such analyses. The following material, including the flowchart shown in Figure B-1, is extracted from the draft guidelines:

The purpose of these guidelines is to develop some consistency in handling the earthquake analyses and design among the various Federal agencies involved in the planning, design, construction, operation, maintenance, and regulation of dams. They are intended to be used as general guides and are not to be considered as standards. It is recognized that the various agencies have differences in mission and diversified location which make agency independence desirable. It is further recognized that earthquake engineering is in the developmental stage and flexibility is desirable. While the content of these guidelines generally reflects current practices, it will be necessary to make periodic revisions, additions, deletions, etc., to maintain currency with the state of the art in earthquake engineering.

When the evaluation of the earthquake factor is completed, the maximum design earthquake (MDE) and the operating basis earthquake (OBE) are selected on the basis of an integrated evaluation of the earthquake factors. The MDE is the largest earthquake used in the seismic analysis of the dam and is generally equated to the controlling maximum credible earthquake (MCE) for the site. The OBE, usually smaller than the MDE, represents the maximum level of ground shaking that can be expected to occur at the site during the economic life of the dam. It may not be possible to show that all possible tectonic features have been discovered. Based on investigations,

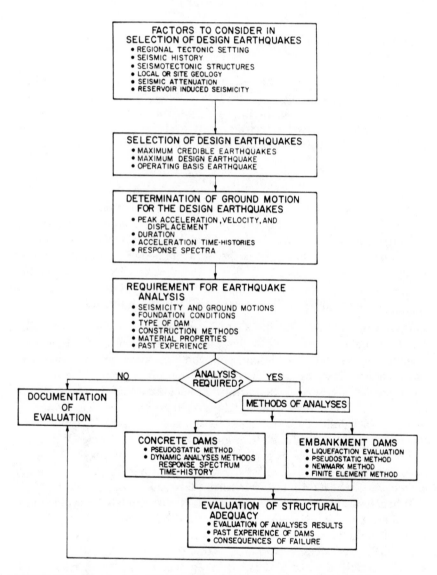

FIGURE B-1 Flowchart depicting steps for earthquake analyses and design of dams. Source: ICODS draft of proposed guidelines.

gaps of information may exist. If so, conservatism may be desirable dependent upon the potential hazards associated with the dam.

1. *Maximum Credible Earthquakes*

The first part of the investigation for selecting the MDE is to estimate the hypothetical MCE for each potential earthquake source, judged to have a significant influence on the site, from the information developed in section D. The hypothetical MCE for each seismotectonic structure or source area within the region examined is defined by magnitude and/or intensity, epicentral distance and focal depth. These MCEs are candidates for the controlling MCE.

2. *Controlling Maximum Credible Earthquake*

The second part of the investigation is to select the controlling MCE for the site as follows:

a. Select the most conservative distance from each seismic source to the site.

b. For each candidate MCE select strong motion records of earthquakes which have similar source and propagation path properties and were recorded on a foundation similar to that of the structure or, if these site-matched records are not available, attenuate the epicentral ground motion parameters or MM intensity to the site using one or more applicable attenuation relationships.

c. Select the controlling MCE based on the most severe ground motion parameters estimated for the site. There may be more than one controlling MCE because of the frequency characteristics on the dam and its components.

3. *Maximum Design Earthquake*

The final selection of the MDE considers whether or not the dam must be capable of resisting the controlling MCE, which is a "worst case" situation. Usually, the MDE is equated with the controlling MCE. However, where the failure of the dam presents no hazard to life, a lesser earthquake for the MDE may be justified providing there are cost benefits and the risk of property damage is acceptable.

4. *Operating Basis Earthquake*

The second level of design earthquake, the OBE, represents the maximum level of ground shaking that can be expected to occur at the site during the economic life of the project, usually 100 years for dams. It reflects the desired level of protection for the project from earthquake-induced structural and mechanical damage and loss of service during the project's economic life, or remaining economic life for existing dams.

The OBE should be based on a probabilistic analysis which accounts for the time element involved in the definition of the OBE. A probabilistic analysis involves developing a magnitude-frequency or epicentral intensity-frequency (recurrence) relationship for each seismic source; projecting the recurrence information from regional information and past data into forecasts concerning future occurrence; attenuating the severity parameter, usually either peak ground acceleration or MM intensity, to the site; determining the controlling recurrence relationship for the site; and finally, selecting the design level of earthquake based upon an acceptable probability of exceedance and the project's exposure period selected for the design.

Soil Conservation Service, U.S. Department of Agriculture
(From letter dated May 21, 1984, criteria presented in Technical Release No. 60, revised August 1981)

SCS has established three classes of dams as follows:

Class (a)   Dams located in rural or agricultural areas where failure may damage farm buildings, agricultural land, or township and country roads.

Class (b)   Dams located in predominantly rural or agricultural areas where failure may damage isolated homes, main highways or minor railroads or cause interruption of use or service of relatively important public utilities.

Class (c)   Dams located where failure may cause loss of life, serious damage to homes, industrial and commercial buildings, important public utilities, main highways, or railroads.

Technical Release No. 60 contains the following requirement:

Seismic Assessment—Dams in zones 3 and 4, Alaska, Puerto Rico and the Virgin Islands and high hazard (class c) dams in zone 2 require special investigations to determine liquefaction potential of noncohesive strata, including very thick layers, and the presence at the site of any faults active in Holocene time. As part of this investigation, a map is to be prepared showing the location and intensity or magnitude of all intensity V or magnitude 4 or greater earthquakes of record, and any historically active faults, within a one-hundred kilometer radius of the site. (Obtain earthquake information for this map in print-out form from the Environmental Data Service, attention D62, NOAA, Boulder, Colorado 80302. Telephone: FTS 323-6472; Commercial (303) 499-1000, Ext. 6472.) The report should also summarize other possible earthquake hazards such as ground compaction, landslides, excessive shaking of unconsolidated soils, seiches, and in coastal areas, tsunamis.

T.R. No. 60 contains two Seismic Zone Maps, one for the contiguous United States and one for Hawaii, that are labeled as being adapted from TM 5-809-10, April 1973. On each of these maps is shown the following table of minimum seismic coefficients:

Zone	Coefficients
0	0.00
1	0.05
2	0.10
3&4	Base on Seismic Assessment

Elsewhere in T.R. No. 60 the seismic coefficient is defined as "the fraction of a weight to be used as a horizontal force in a quasistatic analysis." Also, the minimum factors of safety for embankment slope stability with seismic forces are given as:

Structure Classification	Minimum Factor of Safety
(a)	1.0
(b)	1.1
(c)	1.1

Tennessee Valley Authority
(From letter dated June 7, 1984)

The General Manager of TVA has pointed out that only a few of the older TVA dams were analyzed for earthquake loadings when they were designed but state-of-the-art methods have been used in the design of newer dams and in analysis of the older dams. The information supplied indicates these methods include dynamic analyses, finite element modelling and evaluations of liquefaction potentials. The following is extracted from a technical paper outlining TVA practices prepared by members of the TVA staff:

Earthquake evaluation of dams is, and will continue to be, in the developmental stage as more becomes known about earthquakes and their effects on dams. Therefore, flexibility and sound engineering judgment are essential in the evaluation to reflect the specific conditions of each dam.

The TVA region is located in an area of low to moderate earthquake activity. Therefore, the risk of large earthquakes occurring and affecting

TVA dams is very low. Some of the TVA dams were originally designed for earthquakes and some were not designed for earthquakes. TVA is presently making an earthquake evaluation of all of its dams.

TVA dams fall into two categories—earthfill and concrete gravity structures. The earthfill dams are mainly rolled compacted filled dams, but there are a few hydraulic filled earth dams and also rockfilled dams.

The earthquake evaluation of TVA dams is based on the following factors:

1. Geological and seismological evaluation of the dam site area and surrounding region to determine the design earthquakes and their ground motions.
2. Material properties of the dam and foundation.
3. The need for earthquake analyses based on the magnitude of the design earthquakes, type of dam, risk and consequences of failure, and past experiences of similar dams shaken by earthquakes.
4. The type of earthquake analyses to be performed.
5. Evaluation of the earthquake analysis results to determine the structural adequacy and safety of the dam.

The evaluation of dams considers two levels of design earthquakes: the maximum credible earthquake (MCE) and the operating basis earthquake (OBE). The MCE is defined as the earthquake associated with specific seismotectonic structures, source areas, or provinces that would cause the most severe vibratory ground motion or foundation dislocation capable of being produced at the site under the currently known tectonic framework. The OBE is defined as the earthquake for which the dam is designed to resist and remain operational. The OBE is usually determined on a probabilistic basis considering the regional and local geology and seismology and reflects the level of earthquake protection desired for operational or economic reasons.

Conclusions

The risk of earthquake damage to TVA dams is very low because of the low to moderate earthquake activity in the TVA region. The earthquake evaluation of the dams is based on geological and seismological studies, field and laboratory tests (to determine material properties), analysis, and engineering judgment. The analysis begins with the simplest methods and conservative assumptions and progresses to more thorough analysis as required. The final evaluation of the seismic safety of the dam is based on all pertinent factors involved and not just the numerical results of the analysis.

U.S. Army Corps of Engineers (For Corps Projects)
(From May 24, 1984, letter, Criteria are set forth in ER 1110-2-1806, dated May 16, 1983)

Criteria used by the Corps for projects designed by the Corps have been described as follows:

a. Earthen dams. Earthen dams are to be capable of retaining the reservoir under conditions induced by the maximum credible earthquake (MCE). Deformation (under MCE motions) is acceptable provided such deformation would not result in loss of the reservoir. The MCE is defined as the earthquake that would cause the most severe vibratory ground motion or foundation dislocation capable of being produced at the site under the currently known tectonic framework. It is determined by judgment based on all known regional and local geological and seismological data. The procedure used in determining the maximum earthquake is deterministic. Where historically based recurrence intervals are used in the determination of the MCE, the selection is made at a recurrence interval which is believed to represent the largest earthquake possible under known tectonic conditions.

b. Concrete dams. Concrete dams are also to be capable of retaining the reservoir under conditions induced by the MCE. Inelastic behavior with associated damage is permissible under the MCE. In addition, concrete dams must resist an operating basis earthquake (OBE). The OBE represents the maximum level of vibratory ground motion that can be expected to occur at the site during the economic life of the project, usually 100 years. The OBE is generally more moderate than the MCE. It reflects the desired level of protection for the project from earthquake-induced structural and mechanical damage and loss of service during the project's economic life, or remaining economic life in the case of existing dams. The OBE is determined by probabilistic methods.

c. Dynamic analyses for earthquake deformations are generally made under selected reservoir loading conditions which are more severe than normal operating conditions but do not represent maximum flood pool. Selected impoundment levels are those which are judged likely to exist coincident with the selected design earthquake event.

d. Because of limited available resources, the Corps also uses a decision process in determining which existing projects having earthquake-related deficiencies should receive priority action. Some of the factors used in the decision process for hydrologic-related deficient projects also apply to earthquake-related deficient projects. An additional factor specific to earthquake-related deficient projects is the high potential for seismic activity.

ER 1110-2-1806 provides for use of seismic coefficient method for determining the sliding and overturning stabilities of concrete structures with coefficients as follows:

Zone	Coefficient
0	0.00
1	0.05
2	0.10
3	0.15
4	0.20

ER 1110-2-1806 also requires a dynamic response-type of analysis for concrete structures in Zones 3 and 4 and in Zone 2 when the site-specific peak ground acceleration for the design earthquake is 0.15g or greater.

For evaluating the seismic response of embankments and soil foundations, ER 1110-2-1806 rules out the seismic coefficient or pseudostatic method and requires analytical techniques to evaluate liquefaction potentials and to estimate deformations. Such analyses are required for all projects in Seismic Zones 3 and 4 and for those projects in Zone 2 where susceptibility to liquefaction or excessive deformation is suspected.

U.S. Army Corps of Engineers
(For National Dam Inspection Program)
(From ER 1110-2-106, dated September 26, 1979)

The "Recommended Guidelines for the Safety Inspection of Dams" issued by the Corps for use in the inspection of non-Corps dams that was authorized by P.L. 92-367, provided for two levels of investigative effort. Phase I investigations are to identify expeditiously those dams which pose hazards to human life and property. Phase II investigations are to evaluate safety of those dams for which Phase I investigations indicate additional in-depth studies are needed.

The guidelines for Phase I investigations call for assessment of potential vulnerability to seismic events based on location of project within the various zones of seismic activity, type of dam, local geology, etc. The guidelines for Phase II investigations call for more elaborate analyses ranging from the conventional equivalent static force or pseudostatic method to "state-of-the-art" investigations and analyses for important high hazard projects.

U.S. Nuclear Regulatory Commission (NRC)
(From letter dated June 8, 1984)

The following is extracted from material furnished by the NRC:

Although the Nuclear Regulatory Commission (NRC) by itself does not plan, design, construct or operate dams, the NRC does regulate dams whose failure could result in a radiological risk to the public health and safety. By virtue of this regulatory responsibility, which is described in the Code of Federal Regulations, the NRC has developed guidelines and design criteria for addressing flood and earthquake hazards, which applicants for permits and licenses to operate nuclear facilities are required to meet.

The regulations and criteria are primarily related to the design and construction of nuclear power plant structures, systems, and components. The Nuclear Regulatory Commission is also involved with the regulation of embankment retention systems for uranium mill tailings where the radiological risk to the public health and safety is considerably less than it is with nuclear power plants. In recognition of this reduced risk, less stringent flooding and earthquake design criteria have been considered for special site conditions (small dams built in isolated areas), where the dam failure would neither jeopardize human life nor create damage to property or the environment beyond the sponsor's legal liabilities and financial capabilities.

Definitions

The "Safe Shutdown Earthquake" is that earthquake which is based upon an evaluation of the maximum earthquake potential considering the regional and local geology and seismology and specific characteristics of local subsurface material. It is that earthquake which produces the maximum vibratory ground motion for which certain structures, systems, and components are designed to remain functional. These structures, systems, and components are those necessary to assure:

(1) The integrity of the reactor coolant pressure boundary.

(2) The capability to shut down the reactor and maintain it in a safe shutdown condition, or

(3) The capability to prevent or mitigate the consequences of accidents which could result in potential offsite exposures comparable to the guideline exposures of this part.

The "Operating Basis Earthquake" is that earthquake which, considering the regional and local geology and seismology and specific characteristics of local subsurface material, could reasonably be expected to affect the plant site during the operating life of the plant; it is that earthquake which produces the vibratory ground motion for which those features of the nuclear power plant necessary for continued operation without undue risk to the health and safety of the public are designed to remain functional.

A "capable fault" is a fault which has exhibited one or more of the following characteristics:

(1) Movement at or near the ground surface at least once within the past 35,000 years or movement of a recurring nature within the past 500,000 years.

(2) Macro-seismicity instrumentally determined with records of sufficient precision to demonstrate a direct relationship with the fault.

(3) A structural relationship to a capable fault according to characteristics (1) or (2) of this paragraph such that movement on one could be reasonably expected to be accompanied by movement on the other.

For detailed discussion and applicable guidelines for seismic analysis and design of uranium mill tailing dams the NRC Regulatory Guide refers to ER 110-2-1806 of the U.S. Army Corps of Engineers.

PART 2—STATE AGENCIES RESPONSIBLE FOR DAM SAFETY

Alaska
(From letter dated May 16, 1984)

State uses criteria consistent with Corps of Engineers Seismic Zone map for Alaska with requirements based, also, on size of the structure, impoundment volume, and hazard class.

Arizona
(From letter dated June 7, 1984)

The following is extracted from material supplied by the Chief, Division of Safety of Dams, Arizona Department of Water Resources:

The Department does not have formularized criteria or standards for earthquake-resistant design of dams. The assessment of the earthquake hazard, design parameters required, and special design details that might be deemed necessary are determined on a site-specific basis.

In general, ADWR requires that the earthquake hazard assessment for each dam include consideration of the following:

1. Regional tectonic setting.

2. Seismic history of the area within a minimum radius of 100 miles of the dam.

3. Evaluation of potentially active faults within a minimum radius of 100 miles of the dam.

4. Review of seismic zoning maps.

5. Estimation of design earthquake.

Design parameters are assumed based on the earthquake hazard assessment. In most cases a slope stability analysis is required; for most dams the pseudostatic method of analysis is satisfactory but for unusual conditions a dynamic analysis could be required. If conditions warrant, special earthquake-resistant design measures, based on standard engineering practice, are required.

The extent of the assessment of the earthquake hazard, the degree of the conservatism of the design parameters assumed, and the type of any special design measure required depend on the purpose and method of operation of the dam, the size of the dam, and the downstream hazard. These are determined by engineering judgment.

Arkansas
(From letter dated May 14, 1984)

State has no design standards with respect to earthquake hazards.

California
(From letter dated June 1, 1984)

The following is quoted from letter of Chief, Division of Safety of Dams:

In response to your May 4, 1984 letter, we are outlining our approach to hydrologic and earthquake related safety criteria and standards. As a matter of policy we do not publish standards or criteria so the information provided herein has been compiled from several internal documents specifically to answer your letter.

Maximum credible earthquakes have been used to assess the seismic stability of essentially all dams evaluated in the last 12 years. The earthquake sources are both active or potentially active faults. Active faults are those which are reasonably believed to have experienced surface or subsurface offset in Holocene time (11,000 ± years). Potentially active faults are those on which the last known activity occurred in Pleistocene time, but are judged to be in a geologic setting conducive to present-day activity.

New dams are expected to withstand these maximum credible earthquakes without incurring severe damage. Existing dams need only retain their reservoirs.

Colorado
(From letter dated June 6, 1984)

The following is extracted from Colorado's response to the NRC request for information:

A minimum stability factor of safety greater than 1.0 is required for any loading condition.

Pseudostatic analyses (seismic coefficient method) are satisfactory for modern constructed dams having soils which do not build up large pore pressures due to earthquake shaking, nor show more than 15 percent strength loss (usually cohesive soils such as clays, silty clays, sandy clays, or very dense cohesionless soils), based on acceptable deformations due to earthquake shaking and crest acceleration less than 0.75g (Rankine Lecture, H. Bolton Seed, 1979).

Slope deformation analyses (dynamic response methods) are required for moderate and high hazard dams which have cohesionless embankments and/or foundations which are subject to liquefaction and the expected peak bedrock accelerations at the site exceed 0.15.

High and moderate hazard dams must be designed to withstand the earthquake loads based upon an analysis of the historic activity, and "active" faults. Sources of data are Colorado Geologic Survey Bulletin #43, by Kirkham and Rogers, and U.S. Geologic Survey publications. Accelerations can be determined by methods developed by Schnabel and Seed.

Geologic and seismic reports much include studies of faults and fault history, and seismicity.

Defensive design measures shall be incorporated in dams subject to earthquake loading, such as extra freeboard drains, filter materials, larger than normal cores, filters, drains, and zoning.

Seismic design criteria are a fairly new requirement in Colorado and we will benefit greatly from the findings of your committee. One of our problems now is the definition of an "active" or "capable" fault. There does not appear to be a universal definition for dam safety. The Nuclear Regulatory Commission has adopted criteria, but it appears too stringent for application to "small" dams. The other dam building agencies have also adopted criteria for "large" dams which also appear too stringent. Another problem is the lack of definition of "Maximum Credible Earthquakes," as related to "small" dams, and if there should be any differentiation between small and large dams with reference to seismic design criteria.

Georgia
(From letter dated June 5, 1984)

The following is quoted from letter of the manager of Georgia's dam safety program:

Currently, the Rules for Dam Safety do not address the problem of stability due to seismic loading for *existing* high-hazard dams. However, on new high-hazard dams that are proposed, the dams have to be proven stable under seismic loading.

Hawaii
(From letter dated May 31, 1984)

State does not have an authorized dam safety program.

Illinois
(From letter dated May 18, 1984)

Seismic requirements can be summarized as follows:

Seismic Stability Analysis Requirements—Dams located in Seismic Zones 1 and 2 do not require a seismic stability analysis. Dams located in Seismic Zone 3 shall, as a minimum, be analyzed for seismic stability using equivalent static load methods. The minimum seismic coefficient to be used for Zone 3 shall be 0.10. Dynamic analysis methods should also be considered for dams in areas that are apt to experience seismic activity.

Indiana
(From pamphlets supplied by Mr. Edwin B. Vician)

Data supplied do not include any criteria for dams.

Kansas
(From letter dated June 12, 1984)

In Kansas, we are primarily concerned with the effects of extreme rainfall amounts on our system of dams and reservoirs. Minimum attention is given to potential losses due to earthquakes except for that portion of Kansas which is identified in Seismic Zone No. 2.

Structures classified as significant or high hazard that are located in Seismic Zone No. 2 are required to have additional geology information and soil mechanics analysis. Based upon results of these tests, certain other safeguard design details are incorporated into the structure such as foundation drains, flatter embankment slopes, increased freeboard, zoning and other related criteria.

Louisiana
(Letter dated May 23, 1984)

No seismic design criteria furnished.

Maine
(From letter dated May 10, 1984)

Regulations are to be developed in near future.

Michigan
(From letter dated May 11, 1984)

The Chief of Dam Safety and Lake Engineering, Department of Natural Resources, reports that Michigan has not issued criteria for dams but follows what is considered good engineering practice in reviews of plans and construction. Because earthquake potential is considered to be low, state has not considered earthquake failure potential in accepting designs or inspecting existing dams.

Mississippi
(From letter dated May 23, 1984)

State basically following SCS practice as set out in Technical Release No. 60, "Earth Dams and Reservoirs."

Missouri
(From letter dated May 10, 1984)

Information supplied by the Chief Engineer, Dam and Reservoir Safety Program, shows the state is making changes to rules and regulations regarding dam safety. In both the existing regulations and the proposed modifications, factors of safety of 1.0 under earthquake loading are required for the following:

• Slope stability of Earth and Rock Conventional Dams, steady seepage, full reservoir
• Structural integrity of concrete conventional dams, full or maximum reservoir
• Slope stability of industrial water retention, steady seepage, full reservoir.

In the above, full reservoir means water level is at the water storage elevation.

The existing regulation requires use of earthquake loadings in accord with seismic risk zones used by SCS and U.S. Army Corps of Engineers. In the draft proposed regulation, earthquake loadings are specified by Table B-1. The zone designations on this table refer to seven separate groups of counties,

TABLE B-1   Required Design Acceleration for Earthquake Design (Missouri)

Dam Type	Stage of Construction	Special Conditions	Environmental Class		
			I	II	III
Conventional or industrial	Completed	Two or more dams in series	0.75 PMA[a]	0.5 PMA[a]	0.5 PMA[a]
		Storage × height greater than 30,000[b]	0.75 PMA[a]	0.5 PMA[a]	0.4 PMA[a]
		Storage × height less than 30,000[b]	0.75 PMA[a]	0.5 PMA[a]	0.4 PMA[a]
Industrial starter dam	After starter dam is finished and before final dam is completed	Any	0.5 PMA[a]	0.2 PMA[a]	0.1 PMA[a]
		Any	0.75 PMA[a]	0.5 PMA[a]	0.2 PMA[a]

Zone	PMA[a]	Intensity[c]
A	0.31	IX-X
B	0.28	IX
C	0.26	VIII-IX
D	0.23	VIII
E	0.20	VII-VIII
F	0.17	VII

[a]PMA is Probable Maximum Acceleration of bedrock determined by the zones.
[b]Storage in acre-feet measured at emergency spillway crest elevation and height in feet.
[c]Modified Mercalli Intensity.
NOTE: The "Environmental Class" listings in the Missouri proposed regulations refer to developments in the area downstream from the dam that would be affected by inundation in the event of dam failure. The classes are defined as:

Class I—contains 10 or more permanent dwellings or any public building.

Class II—contains 1 to 9 permanent dwellings or one or more campgrounds with permanent utility services or one or more roads with average daily traffic volume of 300 or more or one or more industrial buildings.

Class III—Everything else.

with Zone A (the area of highest indicated seismic hazard) being four counties in the immediate vicinity of New Madrid, Missouri.

Nebraska
(From letter dated May 30, 1984)

The Nebraska Department of Water Resources has stated:

The hydraulic and earthquake criteria acceptable during reviewing plans and specifications of dams are relatively the same as those used in this region by the Federal Agencies, particularly the Soil Conservation Service, Corps of Engineers and Bureau of Reclamation. Occasionally, some deviations of criteria are necessary based on existing site conditions and these are resolved on a site by site basis.

New Jersey
(From letter dated May 25, 1984)

Since January 1978, state has been using design criteria established for U.S. Army Corps of Engineers National Dam Inspection Program. Proposed state dam safety regulations have been drafted, but this draft contains no requirements relating to earthquake hazards.

New Mexico
(From letter dated June 4, 1984)

The following is extracted from a summary of current practices furnished by the State Engineer:

The State Engineer has not developed a manual of rules and regulations pertaining to the design and construction of dams because each dam is unique and as such must be designed using current good engineering design practices. Each design is submitted to and reviewed by the State Engineer's staff prior to acceptance by the State Engineer.

Following the State Engineer's endorsement of approval on the application and prior to commencement of construction, the owner nominates an engineer registered in New Mexico to supervise construction of the dam. The State Engineer reviews the qualifications of the engineer and if acceptable he issues a letter approving the engineer and setting forth conditions under which he will supervise construction.

It is the practice of the State Engineer to accept designs prepared under standard engineering procedures. The plans and specifications must be prepared by a registered professional engineer in the State of New Mexico. Each design submittal must be accompanied by sufficient engineering, soils and foundation data to show that under the most adverse static condition the structure has a safety factor of 1.5. Where seismic conditions are indicated the appropriate seismic loading is added to that used in evaluating the most adverse static condition and the safety factor for this situation must be greater than 1.0. A liquefaction potential evaluation is required to accompany the seismic stability evaluation.

New York
(From letter dated May 30, 1984)

The following is quoted from letter of the Chief, Dam Safety Section:

With regard to earthquake hazard, we require investigation for seismic events using an appropriate seismic coefficient depending on the seismic zone that the dam is located in. The seismic coefficients vary from 0.025 to 0.10.

North Carolina
(From letter dated May 18, 1984)

Current state practices have been described as follows:

Regarding earthquakes, our regulations make no specific references to earthquake loading design criteria, but we do have the latitude to require the dam owner's engineer to analyze his dam for earthquake loading on a case-by-case basis. In our region most engineers would apply a .05g to .10g (usually the former) earthquake loading factor in their stability analyses. Rolled fill embankment dams with a steady state slope stability safety factor of 1.5 (our minimum standard) are normally not adversely affected when .05g to .10g earthquake loading is applied in the analyses (i.e., the factor of safety remains above unity, at least in the analyses). However, we would still require earthquake analyses of very high earth dams, for hydraulic fill dams, and for certain concrete dams.

North Dakota
(From letter dated May 29, 1984)

State is in process of developing safety criteria for dam design. No indication of probable criteria for seismic design was provided.

Ohio
(From letter dated June 5, 1984)

The Administrator of the Dam Inspection Section, Ohio Department of Natural Resources, has stated:

As far as designing for earthquakes, we require that normally accepted methods of analysis be employed in assessing stability under such loading conditions. Since most of Ohio is located in seismic Zone I, earthquake-load considerations are not extremely critical to the overall design of a dam.

Pennsylvania
(From letter dated June 5, 1984)

No earthquake design criteria furnished.

South Carolina
(From letter dated May 17, 1984)

The following is quoted from letter of Director, Dams and Reservoirs Safety Division:

Our regulations do not establish specific criteria with respect to earthquake hazards, but instead require that designs be done in accordance with "good engineering practices." In the past, we have interpreted this to mean that earthquakes must be considered by the engineer when he performs stability calculations.

South Dakota
(From letter dated May 16, 1984)

State has no dam safety criteria of its own.

Texas
(From letter dated May 25, 1984)

The following is quoted from letter of the Head, Dam Safety Unit, Texas Department of Water Resources:

We have not yet had occasion to get deeply involved in the question of earthquake criteria for safety of dams. Most of Texas has a "zero zone" rating in seismic probability. In addition, a large part of the dry West Texas zone 1 area (with minor seismic probability) is sparsely settled. A small area of the desert mountains of Big Bend has a zone 2 rating (with moderate seismic probability); even so, there is little opportunity for water development in that region. In summary, it is anticipated that individual review techniques for existing dams, and including plans for new dams, as opposed to a published set of state criteria, will satisfy any needs which Texas may have for earthquake criteria.

Utah
(From letter dated May 8, 1984)

The following was extracted from "Rules and Regulations Governing Dam Safety in Utah," dated January 1982:

Seismic design shall apply to all structures that will be constructed in Seismic Zones U-2, U-3, and U-4 which are classified as High Hazard. The State Engineer may determine that Moderate or Low Hazard structures shall also require seismic analysis.

A map in the Rules and Regulations shows four seismic zones in Utah with Zone U-4 (the most active zone) being an area about 30 miles wide, east to west, extending from the center of the state north to the Idaho border encompassing the areas of Utah Lake and the east part of the Great Salt Lake.

The regulations also set out requirements for seismic studies and analyses, including "determination of the appropriate accelerations associated with the Operating Basis and Maximum Earthquakes" and consideration of such factors as potentials for induced seismicity, creation of seismic waves and induced reservoir instability. Another section specifies that a minimum safety factor of 1.0 is required for an embankment under seismic loading.

Virginia
(From letter dated June 14, 1984)

The "Impounding Structure Regulations" of the State Water Control Board, Commonwealth of Virginia, do not have any specific requirements for consideration of earthquakes in design of dams.

Washington
(Letter dated June 6, 1984)

No seismic design criteria furnished.

West Virginia
(From letter dated May 29, 1984)

The "Dam Control Regulations" issued by the West Virginia Department of Natural Resources specify geotechnical investigations of dam sites, laboratory investigations of foundation and embankment materials and geotechnical evaluations. Two requirements are directed at earthquake loadings: (1) a requirement that embankments have a safety factor of 1.0 under seismic loading and (2) a requirement for consideration of potential for liquefaction.

PART 3—OTHER GOVERNMENTAL AGENCIES

City of Los Angeles, California, Department of Water and Power
(From letter dated July 3, 1984)

The following is adapted from a list of procedures, criteria and standards relating to earthquake hazards furnished by the Department of Water and Power:

1. Analyze dams using maximum credible earthquakes for local and regional events.

2. Goal is satisfactory performance of dams when subjected to the local and regional maximum credible earthquakes.

3. Use two dimensional finite element method and Seed-Idriss approach to analyze the stability of dams.

4. Require removal of alluvial deposits and construction of dam on competent bedrock.

5. Require embankment fill compacted to a minimum relative compaction of 95 % based on DWP's Water System Standard (modification of ASTM D 1557-78, from five to three layers, resulting in 33,750 foot-pound per cubic foot).

6. Provide sufficient crest width (25-30 feet).

7. Provide adequate freeboard (a minimum of seven feet, usually ten feet).

8. Provide relatively flat slopes (3 to 1 upstream and 2-1/2 to 1 downstream or flatter).

9. Provide system of internal drains based on Terzaghi's filter design criteria and cutoffs to control seepage through the embankment and foundation to enhance the dynamic stability.

10. Provide zones in internal drainage system to serve as crack stopper to prevent rapid loss of reservoir water resulting from potential cracks in the dam from seismic shaking.

11. Require grouting of foundation and abutments to seal joint seams, fissures, and voids.

12. Encase all pipeline under the dam and construct seepage cutoffs along the pipes.

13. Provide sufficient blow off capability for smaller reservoirs (below 5,000 acre-feet), capability to blow off half of the water in seven days).

14. Provide surveillance and instrumentation to monitor dam performance. These include surveillance by the Reservoir Surveillance Group, reservoir caretakers and patrolmen. Monitoring facilities and instrumentation include seepage drains, observation wells, tiltmeters, strain gages, movement and settlement points, pore pressure piezometers, seismoscope, strong motion accelerometers, and peak recording accelerometers.

15. Meet the requirements of the California Department of Water Resource, Division of Dams.

East Bay Municipal Utility District (EBMUD), California
(From letter dated May 23, 1984)

The following is quoted from a letter of the General Manager of EBMUD:

The East Bay Municipal Utility District does not have formalized or written criteria and standards applicable to dams. We have attempted to apply current "state-of-the-art" criteria, standards, and procedures in both the design of new facilities and the analysis of existing facilities. These are applied to individual dams in specific site situations and are similar, but not identical, for all dams.

The District has, in most cases, utilized Professor H. Bolton Seed's methods for the application of dynamic loads and the analysis of dams under seismic conditions. The determination of the seismic conditions to be expected is a very vital part of such studies because new data are increasingly available.

New York Power Authority
(From letter dated July 24, 1984)

The following is quoted from information furnished by the New York Power Authority (NYPA):

The Authority is currently proceeding with a seismic reassessment of all its concrete and earth dams. In order to maintain a unified approach to the seismic stability analysis of NYPA earth embankments in the present changing regulatory climate, the following procedure has been adopted:

a) Determine earthquake induced ground motions from an appropriate design earthquake;

b) Review and evaluate existing information and geotechnical data to determine appropriate soil/rock properties and seepage flow;

c) Perform simplified analysis of typical dam cross sections for the prediction of dam deformation.

The NYPA suggested program of studies follows a progression from a relatively simple approach using existing data to a decision point where dam performance will be determined to be either acceptable, mariginal, or unacceptable. More sophisticated sampling and analysis techniques could be carried out if it is required or desirable.

Alternate methods currently in use attempt to determine whether or not a particular soil is susceptible to liquefaction. This includes evaluation of in-situ undrained steady state shear strength by means of laboratory cyclic triaxial tests on undisturbed specimens. A disadvantage to this approach is that it relies upon interpretation of variables involved in the recovery, handling and testing of undisturbed soil samples, and is rather expensive. The approach does not provide direct information on the likelihood that liquefaction will be triggered by a given design earthquake.

Depending upon the results of the simplified evaluation program, a second phase may be required. This phase would consist of a more rigorous analytical procedure because: a) certain parametric output, such as material properties of the soil, or the predicted design earthquake ground motions, proved to be critical; and b) the simplified one and two dimensional analyses did not sufficiently define the problem for certain of these critical input values. The parametric input itself may need further refinement either by expert reevaluation of measured values or by limited field investigations. The results of these analyses, together with the associated uncertainties in both the seismic input and the predicted dynamic response of the earth embankment will stand by themselves as a deterministic evaluation (at some given safety factor) or they can be utilized to evaluate the overall seismic safety of the embankment in probabilistic terms and at various risk levels.

Salt River Project, Arizona
(From letter dated June 25, 1984)

The General Manager of the Salt River Project (SRP) furnished material on a seismic study conducted by SRP on Roosevelt Dam, one of six Bureau of Reclamation Projects operated and maintained by SRP. This study was initiated to develop seismic design criteria (response spectrum) for the Maximum Credible Earthquake at Roosevelt Dam on the Salt River. The report and subsequent addenda emphasize the need for consistent terminology so that the geotechnical consultant, seismologist, analyst and design engineer understand the intent and limitations of design criteria. Paper supplied with the report discussed the problem of pyramiding safety failures in design. Other discussion dwelt on the probabilities of various magnitudes of earthquakes *during the life of the structure.*

Santee Cooper (South Carolina Public Service Authority)
(From letter dated July 12, 1984)

The following is quoted from letter of the President, Santee Cooper:

For the past several years Santee Cooper has been conducting state-of-the-art research on our Pinopolis West Dam in an attempt to determine its reaction to a recurrence of the 1886 Charleston Earthquake. Because of this, we are very interested in the establishment of acceptable risk levels for seismicity as well as for spillway adequacy in Federal dams.

As for our design procedures, criteria and standards for dam safety and inspections, Santee Cooper is licensed by the Federal Energy Regulatory Commission (FERC) and follows FERC guidelines regarding dam safety.

PART 4—PRIVATE FIRMS IN UNITED STATES

Acres American, Inc., Buffalo, New York
(From letter dated July 7, 1984)

The following extracts from a paper furnished by Acres American illustrate the organization's approach to design for earthquake hazards:

Selection of Design Earthquake. The approach to the selection of design earthquake is based on the potential risks involved with the failure of the dam. For smaller dams where the associated risks are small, a design earthquake is selected on the basis of the seismic zone in which the project is located, unless there is clear evidence of a source of seismicity near the project which will affect the project. For relatively large dams, where the risks associated with the failure could be significant (these risks could be economic, financial, loss of life and property, or environmental), design earthquake is selected on the basis of comprehensive seismic and geologic studies. In these cases, the Maximum Credible Earthquake (MCE) is determined using both the probabilistic and the deterministic models. The reason for using both models is apparent lack of complete confidence in any one model at present and the need to assure the safety of the dam. The reason for the lack of confidence lies in the fact that it is not always possible to uncover and study all the earthquake related faults in the project area with a high degree of confidence and the historical seismic records are less than complete for a realistic probability model. Once the MCE events from both models are determined, we use the more severe of the two earthquakes in the design since the increase in cost to provide required safety measures is generally small.

Selection of Design Motion.   Once the maximum credible earthquake, including its source and mechanism, is defined, the key parameters such as maximum free field acceleration, predominant period and duration are determined using currently accepted correlations. The earthquake motion is attenuated to the dam site using appropriate attenuation relationships. In our experience, regional attenuation curves are only available in selected regions and the improvement in results using these curves is still questionable primarily due to the lack of sufficient data for moderate to large earthquakes. For this attenuated earthquake, a response spectrum is determined and appropriate acceleration time history of the earthquake is generated. The time history is generated by either scaling an actual recorded earthquake motion under similar geologic conditions or artificially by using mathematical formulation. In case of actual recorded time history, sometimes an ensemble of time histories is used to represent a wide range of frequency and randomness characteristics.

Analysis of Dam.   The analysis has two distinct elements; the engineering properties of the material and the mathematical calculations. The engineering properties are selected from the published literature and previous project during the preliminary stage of the project and from the field and laboratory tests for the final design. However, difficulties exist in characterizing large size material properties (rockfill, boulders) and Acres, like other organizations, uses published relationships supplemented with parametric sensitivity analysis to study the impact of error in material properties.

The mathematical approach is selected on a project by project basis after considering factors such as dam type and height, reservoir size, type of material in the dam, regulatory requirements and any specific requirements by owner of the facility. Generally, the following approach is used for the earthfill/rockfill dams:

(a) All dams are analyzed using a pseudostatic approach of analysis to study the potential of mass failure. A horizontal seismic coefficient equal to $2/3$ of the maximum peak acceleration is used in the analysis. For dams with significant height, the maximum peak acceleration at various heights within the dam is calculated from response and analysis (either simplified such as Newmark method or more detailed using one-dimensional or two-dimensional model and design time history). A minimum factor of safety of 1.0 or larger is considered acceptable. The soil properties used are those representing the undrained behavior under consolidated stress conditions.

(b) If the dam is constructed of materials which are considered not to experience more than 10 to 15 percent loss of strength due to shaking, a more detailed dynamic analysis is not usually performed. The results may be supplemented by calculating permanent deformations using simple methods such as Newmark analysis and a post-earthquake stability analysis using effective stress parameters and excess pore pressures generated during earthquake. The Newmark type analysis is performed to assist in the development of adequate freeboard.

(c) For materials which may experience significant strength loss due to shaking and for dams, say, higher than 300–400 ft, a more comprehensive dynamic analysis is performed using state-of-the-art techniques such as developed by Seed et al. Sometimes these analyses may be performed even for smaller structures when required by the owner or the regulatory agency. The techniques are well documented in the engineering literature and we maintain a library of software and an analysis group to perform these studies. These analyses may include determination of overall factor of safety based on stress levels, anticipated deformation along potential failure planes or identification of zones that may have potential for tensile stresses. As stated earlier, perhaps the single most important element of difficulty is the determination of the in situ behavior of the coarser soils and rockfill materials, and for this reason alone, we consider these analyses to be good guides and not the specific tools to prepare a safe design.

For the concrete dams, the approach is much more defined. Again, the judgment for the type of analysis is made on a project by project basis. In many cases, a pseudostatic analysis is performed to determine the overall stability and to calculate the stresses. For arch dams and gravity-arch dams, this analysis is supplemented with dynamic response analysis. As a first step, a response spectrum analysis is used to assess the stresses in the structure. If the stresses are within the allowable limits, no further analysis may be performed since the response spectrum analysis yields conservative results. However, if the resulting stresses exceed the allowable stresses, then a time history response analysis is performed using finite-element modelling developed by the USBR which analyzes linear deformation. This analysis is further refined by performing a single cantilever time history analysis incorporating nonlinear springs in beam elements to more accurately estimate tensile stresses in the cantilever.

The results of these analyses are critically reviewed for appropriateness of input data and reasonableness of results.

Design of Structure. The safety of the dam against damage and/or failure during an earthquake is assured through prudent design.

For the earthfill/rockfill dams, the safety of the dam is assured by using appropriate construction materials and procedures and by performing necessary foundation preparations. The use of materials that will dilate during seismic events is encouraged and the upstream zoning is provided with high permeability materials. Any liquefiable soil in the foundation is excavated to found dam on competent material. Liberal use of cohesionless filters and appropriate drainage provisions are made within the body of the dam. The use of liquefiable or otherwise degrading materials during earthquake is avoided in critical zones. Liberal freeboard is provided to accommodate slumping during earthquake and to provide protection against seiches caused by seismic event. The internal zoning and shaping of abutment is done carefully to minimize tensile stresses and a careful quality control is exercised during construction.

In Acres, we believe that it is essential to make provisions for corrective measures should the dam experience damage due to earthquake. Thus, we give a serious consideration to facilities that could be used to evacuate the reservoir (such as low level outlet) and need for galleries to improve, monitor and rehabilitate drainage and grouting.

Similarly, the safety of concrete dams is enhanced by including specific construction details, utilizing higher strength concrete and providing adequate foundation preparation to improve compatibility between the foundation and the superstructure. Further, concrete structures are not normally preferred on sites where a known or potentially active fault is discovered in the foundation.

Conclusions. The rapid advances in the field of seismicity, computational procedures and material testing and study of failure case histories have greatly increased the understanding of behavior of dams during strong earthquakes. However, many of the parameters are still derived from semiempirical relations and the historical records are less than complete. Therefore, the design of a safe dam for earthquake as well as other loads still requires a great deal of engineering judgment and experience, knowledge in material behavior and construction techniques and a sound engineering approach. Any step in formulating a universally acceptable approach would help bring consistency in the profession and make it easy to relate the experience on other similar projects.

Alabama Power Company, Birmingham, Alabama
(From letter dated July 18, 1984)

The following is quoted from letter of Alabama Power Company, which reports it has eleven hydroelectric projects on two river systems within Alabama:

In the area of seismic analysis, our approach is simplified. All of our projects are located in a region of very low seismicity, and we have traditionally used a pseudo static load factor of 0.1g for purposes of analysis. This method of analysis has been supported by our independent consultants and approved by the FERC.

Central Maine Power Company, Augusta, Maine
(From letter July 31, 1984)

Structural analysis sheets prepared by Charles T. Main, Inc. for five hydropower projects were furnished by the Central Maine Power Company.

The analysis sheets indicate that the analyses used the pseudostatic method for accounting for earthquake forces with horizontal forces ranging from 0.05g to 0.10g and, in one instance, a vertical force of 0.03g.

Charles T. Main, Inc., Boston, Massachusetts
(From telephoned response by Thomas Neff, June 15, 1984)

The Charles T. Main organization has used the seismic design criteria of the U.S. Army Corps of Engineers in its work in other countries and has encouraged the use of those criteria in countries that had no or less conservative seismic design criteria.

Duke Power Company, Charlotte, North Carolina
(From letter dated July 19, 1984)

Information supplied by Duke Power Company indicates its standards for dams comprise regulations of the Federal Energy Regulatory Commission supplemented by standards and criteria issued by a number of federal and state agencies.

Planning Research Corporation (PRC), Denver, Colorado
(From letter dated June 19, 1984)

The following is extracted from a description of the design criteria for seismic standards used by PRC:

Whether working in the United States or overseas, we normally define the seismic design parameters by analyzing each specific site. For U.S. projects, we use published design criteria such as the Algermissen Map or building code recommendations as a check, but not as the only method of determining the seismic design parameters.

Our designs are normally based on a site specific Design Basis Earthquake (DBE) which is defined as the largest earthquake which would be expected to occur once during the expected life of the project. We design a project to withstand the DBE with minimal or no possibility of damage occurring and then check to ensure that catastrophic failure will not take place when/if a Maximum Credible Earthquake occurs.

With regard to reservoir induced seismicity (RIS), we normally assess the potential for this phenomenon to occur based on site characteristics, reservoir size, and reservoir depth. A comparison of the particular site under study with projects that have experienced RIS in the past is made and if it is judged that RIS could occur at the site in question, appropriate instruments are incorporated in the design and installed prior to project construction. This monitors the pre and post reservoir filling conditions to determine if, in fact, RIS is experienced and the magnitude of the reservoir induced earthquakes. If the potential for RIS is judged to be low or nonexistant, we usually recommend that a much simpler instrumentation network be installed as a normal part of the overall project instrumentation package in order to monitor the performance of the structure under seismic loading.

R.W. Beck and Associates, Seattle, Washington
(From letter dated June 12, 1984)

The following are extracts from statements of "Design Criteria" and "Methods of Analysis" for an arch dam at Swan Lake Hydroelectric Project, Alaska, supplied by R.W. Beck and Associates to illustrate the firm's practice:

Design Earthquake (DE)

The DE is defined as the largest earthquake that would be expected to occur during the economic life of the dam (recurrence interval of once in 100 years). ("Largest earthquake" implies the earthquake producing the greatest loading on the structure.) The magnitude of this event is determined from magnitude versus frequency of occurrence relationships. The dam is required to safely withstand the loads due to the DE although some repairable damage is acceptable for this loading, if it occurs. Those systems and components important to safety must remain operable.

Maximum Credible Earthquake (MCE)

The MCE is defined as the earthquake that would cause the most severe vibratory ground motion capable of being produced at the site under the currently known tectonic framework. It should be a rational and believable event which can be supported by all known geologic and seismologic data. It is a judgment based on the maximum earthquake that a tectonic region could produce, considering the geologic evidence of past movement and the recorded seismic history of the area. The dam was analyzed to ensure that it can withstand the loads from the MCE without any sudden or uncontrolled release of the reservoir even though damage may occur.

Dynamic Analysis

a. *General*

Because the project site is in a seismically active region, a detailed dynamic analysis of the dam was essential. Two different approaches for calculating stresses due to earthquake loading were used, the response spectrum method and the time-history method. The former gives more approximate and generally conservative answers but is convenient and economical. The time-history method is more complex but has the advantage of showing stresses varying with time. The maximum stress values resulting from each method, however, correspond reasonably well.

A stress analysis of an arch dam acting elastically under earthquake loading is performed as follows:

(1) Define the mass and stiffness characteristics of the dam taking into account the effective hydrodynamic mass considered to move with the structure as it vibrates;

(2) Determine natural frequencies and mode shapes;

(3) Calculate for each mode considered the inertial forces and stresses due to a unit spectral acceleration at each point of the structure;

(4) Determine stresses contributed by each mode based on an actual input ground motion; and

(5) Determine the total stresses.

b. *Response Spectrum Method*

A response spectrum is a plot of the relationship between the maximum response of a series of single degree-of-freedom systems, each having a different period of vibration. The response can be plotted as acceleration, velocity, or displacement. Total stresses are determined by taking the root-mean-square value of the sum of the stresses contributed by each significant mode of vibration.

c. *Time-History Method*

The time-history method differs from the response spectrum in methodology and in that stresses are computed for discrete times during the earthquake.

The stress contributed by each significant mode at a given instant of time is determined and total stress at each mode at a given instant of time is determined by mode superposition. Thus, it is possible to obtain stress distributions throughout the dam at any given instant of time.

Yankee Atomic Electric Company, Framingham, Massachusetts

A company representative has made available a report, dated December 1982, titled "Supplemental Seismic Probability Study—Yankee Atomic Electric Company, Rome, Massachusetts." The report documents extensive studies of the probability of earthquakes affecting the site of a nuclear power station, specifically to develop estimates of earthquake shock effects having annual probabilities of 10^{-3} and 10^{-4}. The report does not discuss design criteria for earthquake hazards as such, but shows the problems in attempting to analyze earthquake potentials in the eastern United States.

APPENDIX **C**

Probable Maximum
Precipitation (PMP) Estimates

A brief account of the historical development of methods of estimating maximum flood potentials is given in Chapter 4. Chapter 5 outlines the steps involved in current procedures for estimating probable maximum precipitation (PMP). In this Appendix the rationale and methods used in that estimating are discussed.

The methods used have been described in many publications, including a series of Hydrometeorological Reports by the U.S. National Weather Service (1943–1984). These cover a fairly wide range in estimates—from those for specific drainages to those covering a large portion of the United States. A World Meteorological Organization (1973) manual summarizes the various techniques for preparing PMP estimates used in the United States and in some other countries.

BASIC DATA

Since the need is average PMP values for drainage basins, maximum observed values of average rainfalls over standard-sized areas provide the most suitable basic data for PMP estimates. These eliminate, for the most part, working with station or point extremes that cannot be adjusted by a common method to areal average values. A joint effort of the Corps of Engineers and the Weather Service has provided and kept current a loose-leaf publication that contains data on the most extreme observed areal storm rainfalls (U.S. Army Corps of Engineers, 1945). The publication now contains information for about 550 storms. Additional, not as detailed, storm

211

data for approximately 300 storms have been assembled by the Hydrometeorological Branch of the Weather Service. For each storm the greatest areal depths are listed for standard-sized areas, usually from 10 square miles up to 20,000 square miles (depending on the area covered by heavy rain) for standard durations usually from 6 up to 72 hours or more. Methods for computing these areal storm rainfalls from amounts measured at gages are described in a manual (World Meteorological Organization, 1969). An important part of these storm studies is to analyze the meteorological features associated with the heavy rainfall to better understand what factors are important. These studies are necessary for other aspects of PMP estimation described further on.

STORM TRANSPOSITION

An obvious question to ask concerning any major storm is: Why could it not have occurred at some other locality? One type of limitation on a particular storm location is differences in terrain. Significant mountain features (i.e., height, orientation, and steepness of slopes) affect the amount of rain and where it is deposited. Thus, transposition of a storm without adjustment, from a mountainous location to a nonmountainous location or vice versa, is not reasonable.

Meteorological factors also limit where storms can be transposed. If other storms, with the same basic atmospheric factors (moisture source, location of high- or low-pressure centers, frontal positions, for example) have been observed in other locations, this is usually a good indicator that the most severe storm of that type can be transposed to other such locations with appropriate adjustments.

A more generalized transposition procedure has been used for tropical storm rainfalls along the Gulf and east coasts of the United States (Schreiner and Riedel, 1978). This gives adjustments to rain depths based on the decrease with distance inland from the coast observed in the major tropical storms.

ADJUSTMENT FOR MOISTURE

Detailed study of major storms has shown that, if all other factors are the same, the magnitude of rain is related to the moisture content of air flowing into the storm location. Briefly, the mass of water vapor that a column of air can hold is related to the surface dew point assuming the column is saturated and the vertical temperature variation is pseudo-adiabatic. These conditions are generally closely fulfilled in major storms. Thus, in estimating maximum rainfall potentials, we have basis to increase the rainfall in a storm by multiplying by the ratio of the moisture associated with maximum observed dew point in the vicinity of the storm to that observed in the moist air flowing into the storm. Generalized maps of maximum dew points have been published (U.S. National Weather Service, 1968). The dew points representative of the inflowing air for major storms are routinely determined for major storms by the Weather Service.

The maximum dew point charts are also used for adjusting storms when they are transposed. That adjustment is the ratio of the maximum moisture in the transposed location to that where the storm occurred, both based on maximum observed dew points.

ENVELOPMENT

Just as some methods of determining a probable maximum flood (PMF) directly from floods used envelopment, similar envelopments are used for PMP. Such envelopment takes into account the strong possibility that we have not experienced equally extreme rainfall-producing parameters in the observed storms, etc., for all area sizes and durations. Another envelopment step is introduced by utilizing smooth lines and gradients for PMP estimates when making generalized PMP estimates for a region. Such smoothness, of course, is not realistic nor attempted in regions where precipitation is strongly controlled by orography.

APPLICATION TO ESTIMATE FOR SPECIFIC BASIN

A brief summary of the determinations of PMP estimates for a specific drainage basin in nonorographic regions is as follows.

• From meteorological studies, decide which storms in storm rainfall (U.S. Army Corps of Engineers, 1945) can be transposed to the drainage basin of interest. (The U.S. Weather Service has made such determinations for many major storms.)

• Transpose each of these storms by multiplying the storm rainfall depth-area-duration data by the ratio of moisture for the maximum observed dew point at the transposed position to that for the dew point of the inflowing air for the actual storm. This single adjustment increases the rainfall for maximum moisture as well as adjusts it for transposition to the drainage.

• Draw smooth curves for rainfall depth versus area and rainfall depth versus duration enveloping the values determined from the transposed storms. The resulting values from the curves corresponding to the size of the drainage basin provide an estimate of PMP for that basin.

This procedure for estimating PMP involves the assumptions (1) that there are a "sufficient" number of extreme storms and (2) that adjustment for moisture and not for any other meteorological factor is sufficient to give an upper limit. (It is assumed the other meteorological factors combined can be looked upon as "efficiency" and that among the transposed observed storms this efficiency factor reached the highest value that can be expected at the site.)

PROCEDURES FOR MOUNTAINOUS REGIONS

We have stated why storm transposition is not reasonable in mountainous (orographic) regions. Thus, it is necessary to include other methods in such regions. Several different techniques have been used for PMP estimates in orographic regions. One of these is based on adjusting the PMP for the nearest nonorographic location for topographic effects. The adjustment could be based on comparison of extreme rainfall of various categories (station and areal for various durations, rainfall frequencies, etc.). Another is to apply an orographic precipitation computation model, which briefly stated, increases observed storm rainfall for winds as well as moisture and adjusts for differences in topographic features (U.S. Weather Bureau, 1961a). Use of this orographic model must be restricted to substantial slopes of reasonably long lateral extent facing inflowing moisture, and to regions where precipitation in the cool season is most important. For the coterminous 48 states there now remain the eastern Appalachian region and its extension into New England where the published generalized PMP charts are for the most part not applicable (Shreiner and Riedel, 1978). In that report this questionable region is stippled;* where the orographic portion of the total rain is small, the regional study will normally provide only estimates of nonorographic or convergence PMP. The user must evaluate the effects of topography on the precipitation process and incorporate these into the final estimate (Miller et al., 1984; Fenn, in press).

GENERALIZED PMP CHARTS

It has been found that more reliable and consistent PMP estimates result from studies giving mapped values for a region. This is particularly the case for orographic regions where discontinuities from drainage to drainage are minimized.

Procedures for determining generalized PMP estimates are basically the

---

*Another stippled region is shown in that report in the foothill or steep slope region of the Continental Divide. This region is covered by a recent generalized PMP study that does take into account the mountains (Miller et al., 1984).

same as those for a specific drainage. One added feature naturally is that regional smoothing results in consistency from drainage to drainage within a particular region.

The National Weather Service has completed detailed meteorological studies and analyses of data leading to generalized PMP estimates for the 50 states and Puerto Rico. Figure C-1 shows the regions covered by individual studies. Table C-1 lists the studies and other information. These generalized PMP estimates have been adopted by all concerned federal agencies.

APPLICATION OF PMP

The listed PMP studies (Table C-1) give average rainfall values for specific sized areas and for standard durations usually in sufficient detail to determine from the greatest, second greatest, etc., to the least 6-hour incremental rainfall that may be expected in a 72-hour PMP storm. Detailed instructions and recommendations for application of PMP for the United States east of the 105th meridian have been published (Hansen et al., 1982). This report covers the spatial distributions within any specified drainage and the sequence of incremental 6-hourly rain depths in the PMP storm. Details of the study will not be repeated here—rather the considerations used will be discussed in general terms that may be helpful for other regions.

SPATIAL DISTRIBUTION

Should a severe rainfall actually have centered on any specific drainage, it can be used to provide a rainfall pattern for the PMP storm. In nonorographic regions the hypothetical elliptical pattern of Hydrometeorological Report (HMR) 52 (Hansen et al., 1982) can be applied. Some constraints in orientations of the pattern, based on meteorological parameters and observed pattern orientations, may be employed. Generally, the same pattern is recommended for the four greatest 6-hour rain depths, and no areal pattern for the remaining 6-hour depths. This is a simplification of patterns observed in major storms—some of which even show holes or depressions in some 6-hour periods.

In mountainous regions no detailed studies have been made on spatial distributions for the PMP. Use of an actual storm pattern, if well centered on the basin, is preferred. Otherwise, the pattern of mean seasonal precipitation or that of the rainfall frequencies (Miller et al., 1973) such as the 100-year rainfall may be useful. Should orographic PMP distribution be given in a National Weather Service report covering the basin, it could also give guidance to the distribution of PMP.

In order to avoid possible excessive maximization, generally the National

TABLE C-1 Generalized PMP Studies for United States

Study	Geographical Region	Scope
Hydrometeorological reports		
No. 36 (U.S. Weather Bureau, 1961a, revision, 1969)	Pacific coast drainage of California	General-storm PMP; areas up to 5,000 mi^2, 6 to 72 h, seasonal values October through April
No. 39 (Schwarz, 1963)	Hawaiian Islands	General-storm PMP, areas up to 200 mi^2, $1/2$ to 24 h
No. 43 (U.S. Weather Bureau, 1966a, addendum 1981)	Columbia River and coastal drainages of Oregon and Washington	General-storm PMP, areas up to 5,000 mi^2 west of Cascades Ridge, areas up to 1,000 mi^2 east of Cascades Ridge, 6 to 72 h, seasonal values October through June. Local-storm PMP east of Cascades Ridge, areas up to 500 mi^2, durations to 6 h, seasonal values May through September
No. 45 (Schwarz and Helfert, 1969)	Tennessee Valley	PMP and TVA precipitation for Tennessee River basins up to 3,000 mi^2 and 72 h
No. 49 (Hansen et al., 1977)	Colorado River and Great Basin drainages. Also provides local storm PMP for all of California	General-storm PMP, areas up to 5,000 mi^2, 6 to 72 h, monthly values. Local-storm PMP, areas up to 500 mi^2, durations up to 6 h, all season values.
No. 51 (Schreiner and Riedel, 1978)	U.S. east of 103rd meridian[a]	PMP from 10 to 20,000 mi^2, 6 to 72 h, all season values
No. 52 (Hansen et al., 1982)	U.S. east of 105th meridian[a]	PMP from 10 to 20,000 mi^2, duration 6 h, all season values (Application report)

No. 53 (Ho and Riedel, 1980)	U.S. east of 103rd meridian[a]	PMP for 10 mi², 6 to 72 h, monthly values
No. 54 (Schwarz and Miller, 1983)	Southeast Alaska	PMP from 10 to 400 mi², 6 to 72 h, seasonal values May through November. Snowmelt criteria provided
No. 55 (Miller et al., 1984b)	U.S. between Continental Divide and 103rd meridian	General-storm PMP, areas 10 to 20,000 mi² in nonorographic regions and 10 to 5,000 mi² in orographic regions, 1 to 72 h, all season values. Local-storm PMP, for selected portions of study region, up to 500 mi², durations 6 h, all season values
Weather Bureau technical papers		
No. 42 (U.S. Weather Bureau, 1961b)	Puerto Rico and the Virgin Islands	General-storm PMP, areas up to 400 mi², 1 to 24 h. Cloudburst PMP, areas to 400 mi², 5 min to 6 h
No. 47 (Miller, 1963)	Alaska[b]	General-storm, areas up to 400 mi², 1 to 24 h

[a]HMR 51, 52, and 53 originally provided PMP for the United States east of the 105th meridian; PMP between the 103rd and 105th meridian from these reports are now superseded by HMR 55. Application portion of HMR 52 is valid for the eastern United States out to the 105th meridian.
[b]This report originally provided PMP values for the entire state of Alaska. PMP estimates from this report for Southeast Alaska have been superseded by HMR 54.

218

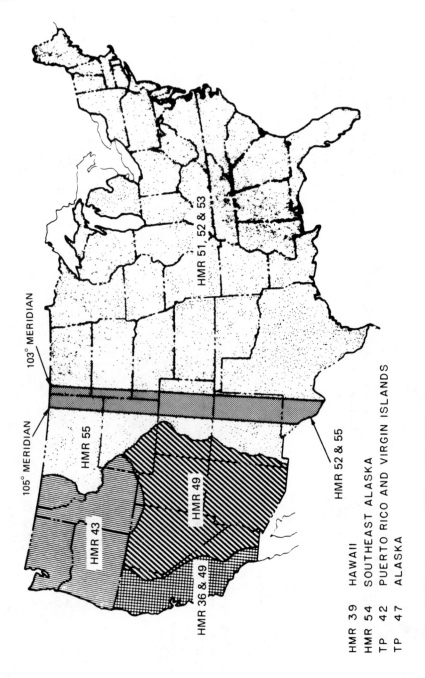

FIGURE C-1 Regions covered by generalized PMP studies.

Weather Service assumes PMP only for the total area of the basin being studied. Thus, less than the PMP is used for smaller areas within the basin and for larger areas including and surrounding the basin. HMR 52 explains in some detail how such specifications are used to determine rainfall within and around the basin.

Generally, it is assumed the PMP storm is centered over the basin of concern—with no travel or movement with time during the PMP storm period. This is a simplification—resulting in possibly a higher or lower PMP and PMF depending on the direction of streamflow relative to direction of possible storm movement. For certain basins it is quite possible that PMP for a smaller area within the basin could be hydrologically more critical than PMP over the entire basin. HMR 52 discusses the rainfall that should be assumed to occur over the residual or remaining part of a basin if the PMP is assumed to cover only part of the basin.

TEMPORAL DISTRIBUTION

Values of PMP are usually given for durations up to 72 hours, unless the basin of concern is small (less than a few hundred square miles), in which case 6- or 12-hour PMP may be adequate. For relatively flat regions, studies have shown that assuming PMP occurs for all durations, out to 72 hours is not an undue maximization.

Sequences of the 6-hour PMP increments (that is, in what order they occur relative to their magnitude) vary widely in major storms. Rain bursts lasting from 12 to 24 hours are most common, although cases can be found with a gradual increase to most intense rain near the middle of the storm. Another consideration is maintenance of PMP for all durations. This can only be obtained if the first and second greatest 6-hour increments are adjacent to each other, and the third adjacent to these two, and so on. With these considerations the following guidance based on storms in mountainous regions is recommended:

• Group the four heaviest 6-hour increments of the 72-hour PMP in a 24-hour sequence, the middle four increments in a 24-hour sequence, and the smallest four increments in a 24-hour sequence.

• Within each of these 24-hour sequences, arrange the four increments in accordance with the sequential requirements. That is, the second highest next to the highest, the third highest adjacent to these, and the fourth highest at either end.

• Arrange the three 24-hour sequences in accordance with the sequential requirements, that is, the second highest 24-hour period next to the highest, with the third at either end. Any of the possible combinations to the three 24-

hour periods is acceptable with the exception of placing the lightest 24-hour period in the middle.

These sequences are similar to those observed in major storms. They also, to a large extent, preserve the basic estimated PMP depths. For example, if the greatest 6-hour increment were followed and preceded by much lesser (smaller) rain increments, the 12-hour depth would be greatly reduced from the estimated PMP value for 12 hours.

SEASONAL VARIATION OF PMP

Certain hydrologic problems require consideration of PMP for seasons other than when the greatest or "all-season" value can occur. For example, PMP for the spring (April or May) in Montana combined with optimum snowpack and melt factors may give a more critical hydrograph than the all-season (most often summer) PMP.

Published studies that include seasonal variation for areas in the United States are listed in Table C-1. The studies, in general, are based on seasonal variations in meteorologic data, including 1,000-millibar dew points (these are an index to atmospheric moisture) and extreme rainfalls of various categories such as maximum station rainfalls of record for several durations, maximum areal rainfalls from storm rainfall (U.S. Army Corps of Engineers, 1945), and climatic summaries. It is believed such precipitation data can be used to extend existing PMP evaluations.

Often when PMP is required for combination with a spring snowmelt situation, optimum sequences of snowmelting parameters (winds, temperatures, dew points, and solar radiation) need to be assessed, as well as an estimate of the optimum snowpack that is available for melt. It is believed that a combination of an estimated maximum probable snowpack, most extreme possible snowmelting parameters, and springtime PMP will give an improbable flood. Judgment needs to be taken to avoid compounding improbabilities. Where spring PMP is the major contributor to the flood for a relatively small basin, a snowpack on the order of a 100-year event would be sufficient. Where snowmelt is the major flood producer, especially for a large basin, at least the greatest snowpack of record (where such records are available) should be used with the PMP. An extreme snowpack is sometimes computed by a hypothetical combination of the greatest recorded precipitation for each of the winter (or snow accumulation season) months. It has been found for several basins in the lower 48 states that observed snowpacks are greater than can be melted by critical sequences of the melt parameters. Thus, something less than this should be used for greatest runoff.

Sequences of snowmelting parameters should be derived from observations either in the basin or adopted from those nearest the basin. A study for a large basin in Alaska (U.S. Weather Bureau, 1966b) used an envelope of highest observed temperatures and winds on the order of 100-year return period values.

EVALUATION OF PMP

It is assumed in a study of PMP, that there is no climatic trend in the meteorologic parameters that would increase or decrease the estimates in the foreseeable future. PMP for the United States is based largely on the greatest recorded rainstorms that have occurred mainly since about 1860 (a few on record prior to this year). As an example consider the 177 storms east of the 105th meridian that have observed rainfall depths that are 50 percent or more of the PMP where the storm occurs (Riedel and Schreiner, 1980) for at least one area size of 10, 200, 1,000, 5,000, 10,000, and 20,000 square miles) and at least one duration (of 6, 12, 24, 48, and 72 hours). The earliest of these storms occurred in 1819, and the latest in 1979. For the decade 1930–1939 there are 30 storms and for the decade 1940–1949, 32 storms. The record shows 15 storms for 1950–1959 and 12 for 1960–1969. At first glance, one could think there may be a climatic trend in the number of extreme rainfalls. This is most likely not the case—the increased data base and interest begun in the 1930s and 1940s are important factors. In addition, the number of storms analyzed is partly dependent on the number of projects and where they are located. Experience with extreme point or station rainfalls (Jennings, 1952) also has not shown a significant climatic trend in their number or magnitude. Users of generalized PMP estimates like to know how rare the values are. This question is difficult to answer, since rainfall data cover at most 100 years. What extrapolations have been attempted indicate the PMP at certain stations range around a return period of millions of years. Relatively little confidence can be placed to such extrapolations.

A comparison has been published (Riedel and Schreiner, 1980) of point (station) 100-year rainfall and 10-square-mile PMP for selected durations. Figures C-2 and C-3 are examples of these comparisons for 24-hour duration. These comparisons show ratios of PMP to 100-year rainfalls ranging from 6 to 2. The highest ratios tend to be in the drier desert regions, the lowest being along mountain ridges. Such variations can be expected, since the more extreme the rainfall, the less the regional range due to topography and other factors. Thus, there is a greater range in the 100-year events than in PMP, and necessarily, there is a regional range in the PMP to 100-year ratio. This ratio, however, is not a simple measure of the rarity of the PMP.

Some data are given in Appendix A on storm rainfalls that exceeded 50 percent of PMP. Figure C-4 shows a map of the storms that had rainfall

FIGURE C-2   Ratios of estimated PMP for 10 square miles to estimated 100-year frequency rainfalls (both for 24-hour durations), eastern United States. Source: Riedel and Schreiner (1980).

depths for 10 square miles and 24 hours that were 50 percent or more of PMP (Ho and Riedel, 1980). There are 84 storm depths exceeding 50 percent for this area size and duration for the region east of the 105th meridan and that west of the Continental Divide. One may be concerned that there are relatively few cases in Nevada, Arizona, and western Oregon and Washington. It is believed this is partly due to much fewer rainfall observations in these regions than, say, in California. Another factor may help explain the distribution shown on the map: the PMP is a rarer event in some regions compared with others. It seems reasonable to expect a relatively rare storm more frequently on the Sierra slopes than in the deserts of Nevada and Arizona.

In an earlier section, several assumptions were listed that are inherent in the procedure for estimating PMP. One assumes a "sufficient" storm sample. If this is not met (a subjective decision) and if no other compensating procedure is applied, it would indicate the PMP estimate is on the low side.

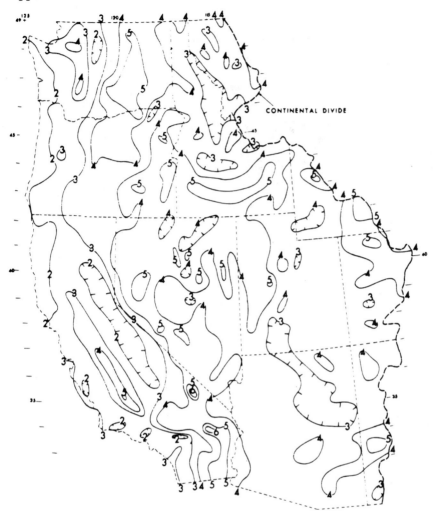

FIGURE C-3 Ratios of estimated PMP for 10 square miles to estimated 100-year frequency rainfalls (both for 24-hour durations), western United States. Source: Riedel and Schreiner (1980).

The other assumption, that adjustment only for moisture is sufficient to obtain an upper limit for storm rainfall, also probably results in a low PMP estimate. However, a possible compensating factor is that the highest "efficiency" (as indicated by extreme observed rainfall) when combined with maximum moisture *may not* yield the most extreme rainfall or the upper limit. Not enough is known about the "best" values of the factors for maximum rainfall.

224

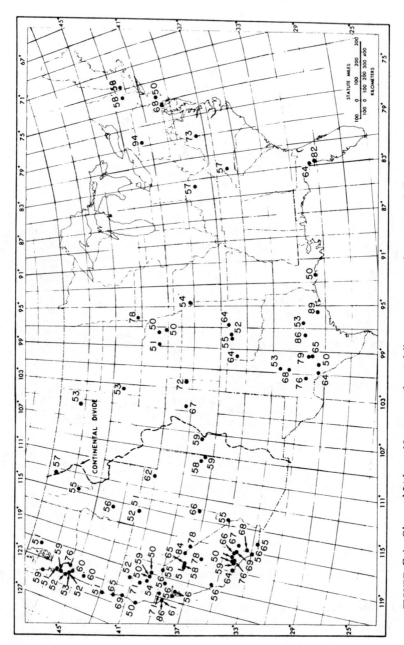

FIGURE C-4 Observed 24-hour, 10-square-mile rainfall amounts expressed as percent of all-season PMP estimates. Source: Riedel and Schreiner (1980).

One bit of evidence lends credence to the assumption that current generalized PMP estimates are on the low side. Generalized studies, covering the same region, if of the same degree of detail and endeavor, over a period of years show an increase with time. Generalized studies for the United States east of the 105th meridian made over a period of 31 years most nearly meet these conditions (U.S. Weather Bureau, 1947; Riedel et al., 1956; Schreiner and Riedel, 1978). Figure 5-1 (see Chapter 5) shows mapped isolines of the ratios of the PMP values for the 1978 study to the PMP values of the 1947 study. The PMP is for 200-square-mile areas for a duration of 24 hours. The map indicates a maximum ratio of 1.37—with considerable areas with ratios above 1.20. These changes in PMP estimates over a 31-year period may be considered moderate, in light of the nature of the estimating process, and on that basis, the PMP estimates may be considered fairly stable. However, one conclusion can be made: that the assumptions on "sufficiency" and "moisture maximization" have not quite stood the test of time. Another way of putting it is that in the future there will be greater rainfalls (for some area sizes and durations) at least in some portions of this particular region. This conclusion includes the fact that lesser storms of certain critical types with time will occur over wider "bounds"; thus, the transposition limits of some most extreme storms will be increased. These two factors are most responsible for the increase in PMP with time.

The following equation (Hershfield, 1961) is sometimes used as a check on PMP estimates or as the basis of design rainfall:

$$X_m = X_n + K_m S_n,$$

where X_m is maximum observed rainfall, X_n is mean of the series of n annual maxima, S_n is standard deviation of the series, and K_m is number of standard deviations needed to obtain X_m.

Using many series of station maximum annual 24-hour rainfall events, mostly in the United States but also some in other countries, the greatest value found of K_m to obtain X_m was 15. Use of this value was suggested to compute statistical PMP. In a later paper (Hershfield, 1965), K was found to be correlated with the raininess of the region, the drier locations having a higher K.

A map covering the eastern two-thirds of the United States (Hershfield, 1961) shows isolines of statistical PMP using a K of 15. It was shown (Riedel, 1977) that this "statistical PMP" has been exceeded by at least 10 observed 10-square-mile storm depths (U.S. Army Corps of Engineers, 1945). While the later adjustment (Hershfield, 1965) was not used in the comparison, which could possibly raise the K value to a maximum of 20, casual inspection indicates these observed 10-square-mile depths would still exceed the statistical PMP. Numerous other comparisons have been made between the deterministic and this statistical PMP. One was for the southwest states region (Hansen et al., 1977). Here the variable K factor was used to compute the statistical PMP for 98 stations. Only 2 of the 98 statistical PMP values exceeded the PMP of the report. Overall, the statistical PMP averaged approximately two-thirds of the report PMP.

From the several methods of evaluation, it can be concluded that the PMP, although extremely rare, is probably on the low side with respect to its definition.

Concepts of Probability in Hydrology

Rainfall depths for specific durations and streamflow peaks occurring during long periods of time are stochastic variables and can be analyzed as such. Statistical and probabilistic analyses allow the development of probability statements or estimates related to the magnitude of certain events. Such estimates can be used for design purposes.

Random or stochastic variables may be discrete or continuous. Rainfall accumulation and streamflow are generally considered continuous, since they may take any value on the real axis (or at least at any positive value). To quantify or parameterize the probability of occurrence of a continuous random variable, one can use a standard probability density function (pdf) $f(X)$. The probability of a random variable, X, taking a value in the infinitesimal range $[X, X + dX]$, is $f(X) \, dX$. Given the pdf $f(X)$, then the probability that the random variable X assumes a value between x_1 and x_2 is

$$P[x_1 \leq X \leq x_2] = \int_{x_1}^{x_2} f(x) \, dx$$

The distributions of a random variable are frequently characterized by shape measures or moments. The mean, a measure of central tendency, is defined by

$$\mu = \int_{-\infty}^{\infty} x f(x) \, dx = E(X)$$

227

where E is called the expectation operator. The variance, a measure of dispersion, is

$$\sigma^2 = \int_{-\infty}^{\infty} (x - \mu)^2 f(x) \, dx = E(x - \mu)^2$$

The standard deviation is the square root of the variance. The skewness is the third moment around the mean, and measures deviations from symmetry:

$$G = \int_{-\infty}^{\infty} (x - \mu)^3 f(x) \, dx$$

Higher order moments are also defined. A standardized measure of asymmetry is the coefficient of skewness:

$$\gamma = \frac{G}{\sigma^3}$$

BULLETIN 17 PROCEDURES FOR FLOOD-FREQUENCY ANALYSIS

Two broad classes of flood-frequency analyses can be identified, depending upon whether or not stream-gaging records exist at or near the location of interest. Procedures available for use at ungaged sites, however, either use directly, or depend on, results of procedures designed for use at gaged sites; they offer no useful insights that are not given more clearly by discussion of procedures designed for use at gaged sites. For this reason, only those procedures designed for use with gage data are discussed here.

In order to promote correct and consistent application of statistical flood-frequency techniques by the many private, local, state, and federal agencies having responsibilities for water resource management, the U.S. Water Resources Council formulated a set of guidelines for flood-frequency analysis known as Bulletin 17, or the WRC procedure (Interagency Advisory Committee on Water Data, 1982). These guidelines, originally issued in 1976, reflect an evaluation, selection, and synthesis of widely known and generally recognized methods for flood-frequency analysis. The guidelines prescribe a particular procedure but do permit properly documented and supported departures from the procedure in cases where other approaches may be more appropriate. Federal agencies are expected to follow these guidelines, and nonfederal organizations are encouraged to do so.

In broad outline, Bulletin 17 characterizes flood occurrence at any location as a sequence of annual events or trials. The occurrence of multiple trials (floods) per year is not addressed. Each annual trial is characterized by a single *magnitude* (peak flood discharge). The magnitudes are assumed to be mutually independent random variables following a log-Pearson Type III probability distribution; that is, the logarithms of the peak flows follow a Pearson Type III (gamma) distribution. This distribution defines the probability that any individual annual peak will exceed any specified magnitude. Given these annual exceedance probabilities, the probabilities of multiyear events, such as the occurrence of no exceedances during a specified design period, can be calculated as explained below. By considering only annual events, Bulletin 17 reduces the flood-frequency problem to the classical statistical problem of estimating the log-Pearson probability curve using a random sample consisting of the record of annual peak flows at a site.

The Bulletin 17 procedures utilize three broad categories of data: systematic records, historical records, and regional information. The systematic record is the product of one or more programs of systematic streamflow gaging at a site. The essential criterion is that the annual peak be observed or estimated each year regardless of the magnitude of the peak. Breaks in the systematic record can be ignored and the several segments treated as a single sample, provided that the breaks are unrelated to flood magnitudes. Thus, the systematic record is intended to form a representative and unbiased random sample of the population of all possible annual peaks at the site.

In contrast to the systematic record, the historical record consists of annual peaks that would not have been observed except for evidence indicating their unusual magnitude. Flood information derived from newspaper articles, personal recollections, and other historical sources almost inevitably must be of this type. The historical record can be used to supplement the systematic record provided that all flood peaks exceeding some threshold discharge during the historical record period have been recorded.

Regional information is also used to improve the reliability of the Bulletin 17 frequency curve by incorporating information on flood occurrence at nearby sites. The regional information can include a generalized skew coefficient that is used to adjust the station skew for short records. Bulletin 17 also provides procedures for adjustment of short-record frequency curves by means of correlations with longer records at nearby stations and for computation of weighted averages of independent estimates.

The Bulletin 17 frequency analysis overall has six major steps:

1. systematic record analysis;
2. outlier detection and adjustment, including historical adjustment and conditional probability adjustment;
3. generalized skew adjustment;
4. frequency curve ordinate computation;
5. probability plotting position computation; and
6. confidence-limit and expected-probability computation.

Bulletin 17 describes each of these steps in some detail. The key to the procedure in most cases reduces to the use of the sample mean x and sample variance s^2 of the logarithms x of the observed peak flows. These two sample moments, along with a weighted average of the sample skewness coefficient and a regional estimate of the skewness coefficient, generally serve to define the estimated frequency distribution of the floods at a site. Several of the steps above pertain to special techniques and adjustments employed to deal with special cases, unusual flood values, and historical information.

Mixtures of hazards can occur in flood-frequency analysis, for example, when several distinct causes of flooding, such as thunderstorms, hurricanes, snowmelt, or ice-jams, can be identified and give rise to floods of such different magnitudes. In such cases, each type of flood can be analyzed in isolation, producing a conditional probability curve for each type. The total or unconditional probability distribution is constructed by weighting each conditional distribution in proportion to the probability that a flood will be of that type (Benjamin and Cornell, 1970; Vogel and Stedinger, 1984).

FLOOD RECURRENCE INTERVALS AND FLOOD RISK

A widely used means of expressing the magnitude of an annual flood relative to other values is the return period or the probability of exceedance. The 100-year or 1 percent chance flood is defined as that discharge having a 1 percent average probability of being exceeded in any one year; this discharge can be estimated from the probability distribution of annual floods. (Note that the 100-year flood is not a random event like the flood that occurs in a particular year; rather it is a quantile of the flood-frequency distribution.) If the occurrence of an annual flood exceeding the 100-year flood is called an exceedance or a success, and if annual floods are independent of each other, then the probability of an exceedance on the next trial is 1 percent, regardless of whether the present trial resulted in success or failure. The probability that the next trial will fail but the second one succeed is 0.99×0.01. The probability that the next success will be on the third trial is $0.99 \times 0.99 \times 0.01$, on the fourth trial $(0.99)^3 \times 0.01$, and so on. Thus, the most likely time for the next success or exceedance of the 100-year flood is on the next trial. The average time to the next exceedance, however, works out to be 100 trials. This paradox is resolved by noting that the probability distribution of in-

TABLE D-1 Approximate Probabilities That No Flood Exceeds the 100-Year Flood

Period, yr	1	2	10	20	50	100	200
Probability, percent	99	98	90	82	60	40	15

TABLE D-2   Probability (in percent) That Indicated Design Flood Will Be
Exceeded During Specified Planning Periods

Average Return Period of Design Flood (yr)	Length of Planning Period (yrs)				
	1	25	50	100	200
25	4	64	87	98	100
50	2	40	64	87	98
100	1	22	39	63	87
200	0.5	12	22	39	63
1,000	0.1	3	5	10	20
10,000	0.01	0.25	0.5	1	2
100,000	0.001	0.025	0.05	0.1	0.2
1,000,000	0.0001	0.002	0.005	0.01	0.02

terexceedance times is very different from the familiar tightly clustered, symmetrical, bell-shaped normal curve, so it is no surprise that intuition developed on the normal distribution should fail here.

A related problem is determination of the probability that a time interval will contain one or more events. Continuing the example, the probability that a 2-year period will contain at least one exceedance of the 100-year flood is one minus the probability that both years will be free of exceedances, or $1 - (0.99 \times 0.99)$, which is very nearly 0.02. Similarly, the probability that a 3-year period will contain at least one exceedance is $1 - (0.99)^3$, which is very nearly 0.03. The approximate probabilities that various planning periods will be without exceedances of the 100-year flood are listed in Table D-1. Thus, there is only a 15 percent chance that a 200-year period will be free of exceedances of the 100-year flood.

On the other hand, if a family moves into a house at the edge of the 100-year floodplain and plans to stay for 20 years until the children are grown, the chances they move before they get flooded are about 80 percent. Thus, the chances one experiences a flood depend on the annual risk (1 percent in this case) and the length of the planning period or anticipated length of exposure to that risk (either 200 or 20 years in these two examples).

This last observation is particularly relevant to the safety analysis of dams. Dams built or in existence today may continue to function and hold water for centuries. Though it seems impractical to try to plan for such long time periods given the imprecision with which we can foresee the future, certainly society should be concerned with the safe operation of existing and proposed dams over the next 50-100 years. Table D-2 considers four planning periods (25, 50, 100, and 200 years) and several design floods. Specified within the table are the chances each design flood value will be exceeded (one or more times) during each planning period. Certainly a 100-year planning period does not correspond to the situation faced by an individual family that intends to stay in a neighborhood or location for only a few years; however, it is very appropriate for cities, counties, and state and federal governments, which are concerned with society's long-term welfare. Consider the 1,000-year flood listed in Table D-2. While it has only a 0.1 percent chance of occurring in any one year, it has a 10 percent chance of occurring in 100 years. Thus, in situations where dam failure or overtopping is likely to have disastrous consequences, and one wants to ensure that such an event is unlikely to occur over a long period of time, one must design for events with very long return periods.

ESTIMATING THE RETURN PERIOD OF THE PMF

In this section we consider if it is possible to credibly estimate the probability that a probable maximum flood (PMF) estimate will be exceeded, even to within an order of magnitude, through examination of the rainfall-runoff relationship and the probabilities of various events. The following section considers whether statistical approaches are able to provide reliable and credible estimates of flood-frequency curves out into the 10,000- to 1,000,000-year event range.

It would be useful if a reliable and credible estimate of the return period or, equivalently, the exceedance probability of a PMF could be obtained by analysis of rainfall-runoff processes. Thus, one would start with an analysis of the frequency of extreme precipitation, as has been done in some studies (e.g., Harriman Dam—see Appendix A, Yankee Atomic Electric Co.). Then one would need to consider how such precipitation totals would be distributed in time and would interact with winds and antecedent conditions within the basin (both moisture levels in the soil and snowpack in some places). All possible combinations of these factors and the probabilities they would occur jointly must then be determined to arrive at the frequency distribution of extreme floods, as indexed by the surcharge over the dam or the flood volume, maximum discharge rate, or jointly by both.

While the algebra can easily be written to describe these computations, calculating the probability of two conditions arising together (two probable maximum precipitations (PMPs) back-to-back or a PMP on the probable maximum snowpack), neither of which has been experienced separately, seems like a very imprecise activity. Newton (1983) has tried to perform such a calculation, and his analysis illustrates the problems.

Newton indicates that one might consider the probability of a near-PMP storm on a particular watershed to be 10^{-8}, with 10^{-6} defining an upper confidence limit. An antecedent storm of 15–50 percent of the PMP is used to define antecedent moisture conditions. He assigned a probability of 1/130 per year for such a storm and 3/365 for the probability that it would occur in the 3 days preceding the PMP. This yields a probability of (3/365)(1/130) or 6 × 10⁻⁵. Thus, that antecedent condition followed by a PMP rainfall has a probability of

$$(6 \times 10^{-5})(10^{-8}) = 6 \times 10^{-13}$$

Now consider the following issue. Certainly PMP-type storms are more likely to occur in some seasons than others and when regional atmospheric conditions are unstable. Given that meteorologic conditions will be such that a PMP would occur, the conditional probability of a 15–50 percent PMP preceding storm may be only 10^{-3}, or even 10^{-2}. This would yield a probability of this joint event of

$$(10^{-2} \text{ to } 10^{-3})(10^{-8}) = 10^{-10} \text{ to } 10^{-11}$$

rather than 6×10^{-13}.

Problems with defining joint probabilities for extreme events appear at every turn. In some situations they may not matter, if the magnitude of the antecedent storm has little effect on the basin's hydrologic response to the PMP and on anticipated antecedent reservoir levels. However, antecedent conditions could be very important in systems with large natural or man-made storage capacity. In any case, the fact that antecedent conditions may or may not have an effect on the actual performance of a reservoir illustrates an additional difficulty with calculations aimed at assigning probabilities to PMF events. A string of unusually rare events may have a joint probability of occurring of less than 10^{-12}. However, if a near-PMP event with frequently experienced conditions is likely to occur with probability 10^{-6}–10^{-8} and for a modest size reservoir is likely to cause nearly as big an outflow hydrograph as the combination with much lower probability, then the 10^{-12} value is more misleading than useful. In some places in the United States, such as California, the PMP value may be as little as twice the 100-year rainfall; in places where extreme rainfalls are very unusual but possible (such as in extreme northern New England), PMP values may surpass the hundred-year rainfalls by much larger factors. These considerations demonstrate the issues that must be addressed by an analysis that would assign return periods to extreme flood events. One would need to consider the frequency of severe rainfalls from modest storms to the worst observed and up to the PMP. These must be convoluted with the conditional frequency of other conditions (soil moisture, snowpack, flood pool elevations, base flow rates) to arrive at the frequency distribution for spillway release rates from a proposed or existing dam.

Development of estimates of the return period of PMF values would be a valuable activity, and such attempts should be encouraged. In some instances, antecedent conditions may have little effect on the PMF values, and the required calculations will be greatly simplified and can credibly be defended. Certainly, the return period of near PMP values varies greatly (Newton, 1983; Washington State response, Appendix A); thus, one would expect the exceedance probabilities of PMFs to vary widely, particularly when one considers the uncertainty associated with the other extreme meteorological and hydrological conditions that are assumed to pertain. Quantification of these factors could make planning more rational if such quantification were credible.

FREQUENCY ANALYSES FOR RARE FLOODS

Bulletin 17 provides uniform procedures for estimation of floods with modest return periods, generally 100 years or less. Extrapolation much beyond the 100-year flood using flood-frequency relationships based on available 30- to 80-year systematic records is often unwise. First, the sampling error in the 2 or 3 estimating parameters is magnified at these higher return intervals (Kite, 1977). Moreover, the available record provides little indication of the shape of the flood-frequency curve or confirmation of a postulated shape at the extreme return periods of interest here. The paragraphs below consider how those problems might be overcome.

Regionalization

If only 30–80 years of record is available at a single site, then perhaps information at many sites can be combined to obtain a better estimate of the frequency of extreme events.

Wallis (1980) discusses one class of such procedures. They involve defining a standardized (dimensionless) flood-frequency curve for a region that is scaled by an estimate of the mean flow at a site to allow calculation of a flood-frequency curve for that site. Such procedures, called the index flood procedure, were once used by the U.S. Geological Survey before being abandoned because they failed to capture the effect of basin size on the shape of the flood-frequency curve (Benson, 1968).

If one had 40 years of record at 100 gages in a region, they would have 4,000 station-years of data. However, this is still far less than the 10,000 to 1,000,000 station-years of data that would be desirable to reliably and credibly estimate floods with return periods on the order of 10^4–10^6 years.

Still there are other problems. Sometimes the drainage area for one gage falls within the drainage area for another gage, so that the two records do not reflect independent experiences. Even when drainage areas do not overlap, annual floods at different gages are cross correlated, reflecting common regional weather patterns; this too decreases the information content of regional hydrologic networks (Stedinger, 1983b, pp. 507–509). Still other problems are caused by the need to standardize the flow records from individual gages if one attempts to define a standardized (dimensionless) flood-frequency curve for a region (Stedinger, 1983b, pp. 503–507).

Others have advocated Bayesian and empirical Bayesian procedures for estimating flood risk. Stedinger (1983a, pp. 511–522) provides a review of much of this literature. With such approaches, the information about the mean and variance of floods provided by the available at-site flood-flow record can be augmented using regional relationships that describe the variation of those quantities with measurable physiographic and other basic characteristics. Unfortunately, the precision of such regional regression relationships appears to be such that the results are worth less than 20 years of at-site data. Thus, the Bayesian procedures generally discussed result in very modest improvements in the precision of flood-frequency curves for sites with 40 years or more of data. Moreover, these procedures are really designed for flood-frequency estimation within the range of experience, again perhaps the 100-year flood or less.

On balance, statistical analysis of available records of annual flood peaks is unlikely to provide reliable and credible estimates of floods with return periods of the order of 10^4–10^6 years.

Use of Historical Flood Data

Physical evidence and written records of large floods that have occurred in the recent and distant past provide objective evidence of the likelihood and frequency of larger floods beyond that provided by gaged flow records. Techniques are being developed to include recent historical information in a given river basin with the available gaged or systematic annual flood record (Condie and Lee, 1981; Cohn and Stedinger, 1983). Use of 200 years of pregaged-record experience might be worth 100 years of systematic record. This would be useful for many purposes but is still insufficient for ours.

Alternatively, multidisciplinary paleohydrologic techniques that rely on stratigraphic and geomorphic evidence of extraordinary floods (Jarrett and Costa, 1982) have the best potential for illustrating what size floods can occur. This is particularly true when a large area is searched that would provide evidence of extraordinary floods, had they occurred. When such techniques are applicable, they should be used to demonstrate that calculated PMFs are credible and are neither unreasonably larger or small; current paleohydrologic techniques seem better suited for illustrating what is or is not possible than they are for constructing a safety evaluation flood.

Numerical Evaluation of Expected Damages

A numerical estimate of the expected damage costs calculated as part of a risk-cost analysis is generally based upon (1) a proposed flood-frequency curve extended to the PMF and (2) a set of flood damage estimates corresponding to discrete and particular flood flows. From these data, one needs to construct the probability density function $f(q)$ for the flood flows and an estimate of the damage function $D(q)$. The function $D(q)$ is used to interpolate the damages associated with various flood flows between the flood flow values for which the corresponding damages have been estimated by flood routing and actual damage estimates. Once these two functions have been constructed, one can numerically integrate their product $D(q)f(q)$ from q_{min} to the PMF to get an estimate of the expected damages associated with floods in this range. Events smaller than q_{min} are assumed not to cause damages while the damages associated with floods larger than the PMF are neglected. Neglecting those large events should not affect the relative costs of various alternatives if the dam would be overtopped and fail for all designs causing equivalent damages. Each step is discussed below.

Estimating the Probability Density Function

If $F(q)$ is the probability of a flood less than q, then the probability density function $f(q)$ is the first derivative of $F(q)$ with respect to q. The best way to calculate $f(q)$ is to develop an analytical description of the flood-frequency curve yielding an analytical expression for $f(q)$. For example, if $F(q)$ is obtained by a linear extension on log normal paper of the flood-frequency curve through the PMF, then over that range

$$f(q) = \{(2\pi)^{0.5}vq\}^{-1} \exp[-0.5(\ln q - m)^2/v^2]$$

where the values of m and v can be determined from boundary conditions such as

$$1 - F(PMF) = 10^{-4} \text{ or } 10^{-6} \text{ and } 1 - F(q_{100}) = 0.01$$

where q_{100} is the 100-year flood.

If an analytical description of the postulated flood-frequency curve cannot be estimated, then a finite difference formula of at least second order can be used to estimate $f(q)$ at the required points (Hornbeck, 1975, p. 22). This approach, though less accurate and more trouble, can produce satisfactory results.

Estimating the Damage Functions

From flood routing exercises using appropriate reservoir operating policies and surveys of the associated damages, agencies can estimate the damages D_i associated with the values of several large inflows q_i and a particular reservoir and spillway design. These values can be used to develop an estimate of the damages $D(q)$ associated with each value of q over this range. Such a function is required in the estimation of the expected damages.

It is generally appropriate to assume that $D(q)$ is a continuous function of q except at the critical flow q_c, above which the dam would be overtopped and would breach and fail. Thus, it is appropriate to develop an estimate of $D(q)$ for $q \le q_c$ (where a breach does not occur) and one for $q > q_c$. A satisfactory estimate of $D(q)$ over each interval can usually be provided by an interpolating cubic spline with natural end conditions (Ahlberg et al., 1967). A piecewise-linear approximation could also be used but would provide less accurate results.

Numerical Integration

Once an analytical expression for $f(q)$ is determined and an approximation for $D(q)$ constructed, one can develop a precise numerical estimate of the expected damages associated with the design or proposal to which $D(q)$ corresponds. This is done by numerically integrating $D(q)f(q)$.

Because $D(q)$ is discontinuous at q_c while some of its derivatives are most likely discontinuous at the q_i points for which D_i was specified, the product of $D(q)f(q)$ should be integrated separately over each interval q_i to q_{i+1} and the results added. This avoids accuracy problems that those discontinuities might cause with some numerical integration algorithms. Over each subinterval one can then employ at least a second-order numerical quadrature (integration) formula (Hornbeck, 1975, pp. 148–150) with a sufficiently small step size to ensure accurate results. Gould (1973) shows that Simpson's second-order numerical integration formula can yield significantly better estimates of expected damages than the commonly used midpoint procedure with a moderate step size. The midpoint procedure can grossly overestimate the actually expected damages associated with a given damage function and a proposed frequency curve. There is no excuse for not carefully processing damage cost estimates with a proposed flood-frequency curve to accurately determine the associated expected damages. This can be done as described above.

A Simple Example of Expected Damages

In Appendix E, simple flood frequency and damage estimation models are presented along with the results of five cases that showed the expected annual damages depending on the characteristics of the project and the parameters of the models. In this section we indicate how the expected damages were computed. The flood-frequency and damage functions were chosen specifically to admit a closed-form solution that would not require numerical integration, but the basic technique would be the same for other functions.

The peak flow rates into the reservoir are assumed to follow an exponential distribution for high flows, so that for $q \geq 10,000$ cubic feet per second (cfs) the probability that the peak flow in any year is less than q is

$$F(q) = 1 - \exp[-r(q + q_o)]$$

The probability density function is

$$f(q) = dF/dq = r \exp[-r(q + q_o)]$$

The probable maximum flood is assumed to be 120,000 cfs with a return period T of 10^4 or 10^6 years, and the 100-year flood ($F = 0.99$) is assumed to be 20,000 cfs. If the cumulative distribution function F is to fit these two points, then r and q must satisfy

$$1 - 1/T \ = F(120,000) = 1 - \exp[-r(120,000 + q_o)]$$
$$1 - 1/100 = F(\ 20,000) = 1 - \exp[-r(\ 20,000 + q_o)]$$

which transforms to

$$r(120,000 + q_o) = \ln T$$
$$r(\ 20,000 + q_o) = \ln 100$$

yielding

$$r = 10^{-5} \ln(T/100)$$

$$q_o = (\ln 100)/r - 20,000$$

For the two possible values of T one obtains the following parameter values:

	T = 10^4	T = 10^6
r	4.6×10^{-5}	9.2×10^{-5}
q_o	80,000	30,000

This leads to the exceedance probabilities for different values of T shown in Table D-3.

In this example there is a maximum possible damage to downstream property, denoted M. At flows below 10,000 cfs, no damage occurs, while above that value, damages follow an exponential curve with M as the limiting value. If the dam is overtopped and breaks, the surge wave is assumed to cause the maximal downstream damages M, to which we must add the cost

TABLE D-3   Exceedance Probabilities for Different
Values of T

	$P[q \geq x] = 1 - F(q)$	
X	$T = 10^4$	$T = 10^6$
10,000	1.6×10^{-2}	2.5×10^{-2}
20,000	1.0×10^{-2}	1.0×10^{-2}
50,000	2.5×10^{-3}	6.0×10^{-4}
75,000	8.0×10^{-4}	6.0×10^{-5}
100,000	2.5×10^{-4}	6.0×10^{-6}
120,000	1.0×10^{-4}	1.0×10^{-6}

from loss of services and rebuilding the dam. This latter quantity is denoted L. Thus, the damage function is

$$D(q) = O \qquad\qquad q \leq 10,000$$
$$D(q) = M\,1 - \exp[-s(q - 10,000)] \qquad 10,000 \leq q < q_c$$
$$D(q) = M + L \qquad\qquad q_c \leq q \leq 120,000$$

where q_c is the critical inflow above which the dam fails. Note that the parameter s determines at what point damages become a significant fraction of the maximum M. If s is "large" ($> 10^{-4}$), significant damages occur at or below the 100-year flood, while if s is smaller, the relatively high damages only occur with larger inflows.

To compute the average annual damages, $D(q)f(q)$ is integrated from the no-damage flow up to the PMF. In our example, if the flow ever exceeds the PMF, the damage will not depend on the particular dam and spillway design selected. Thus, the expected damages of interest are

$$D = \int_{10,000}^{120,000} D(q)f(q)\,dg$$

$$= \int_{10,000}^{q_c} M\{1 - \exp[s(q - 10,000)]\}\,f(q)\,dg$$

$$+ \int_{q_c}^{120,000} (M + L)f(q)\,dg$$

Notice that the integral has been broken where $D(q)$ is discontinuous. In most cases one would have to perform the integration numerically. However, the probability and damage functions have been selected so that a closed-form expression for the expected annual damages can be obtained. Here the probability density function is

$$f(q) = r \exp[-r(q + q_o)]$$

so that the first integrand becomes

$$D(q)f(q) = Mr \{\exp[-r(q + q_o)] - K \exp[-(r + s)q]\}$$

where

$$K = \exp[10{,}000s - rq_o]$$

The expected damages are then

$$E[D] = M \left\{ -\exp[-r(q + q_o)] + \frac{rK}{r + s} \exp[-(r + s)\, q] \right\} \Big|_{10{,}000}^{q_c}$$

where

$$F(q) = 1 - \exp[-r(q + q_o)]$$

The average annual damages depend on the damage function parameters M, L, and s; the parameters of the flood flow frequency distribution r and q_o (which in turn depend on the return period assigned to the PMF, T); and the critical flow q_c. In general, dam design and operating policy influence the level of damages that will result from different flood events, and these in turn determine the average annual damages.

APPENDIX **E**

Risk Analysis Approach to Dam Safety Evaluations

This Appendix considers the issues involved in decision analysis procedures addressing the consequences of dam operation and dam failure during extraordinary flood events. Issues addressed are extrapolation of the frequency curve to the probable maximum flood (PMF), calculation of the consequences of large floods and the matrix decision approach, calculation and use of expected costs, and finally the justification of doing a careful risk analysis of project operations. An example of the application of risk analysis to design alternatives for a hypothetical reservoir problem is outlined.

EXTRAPOLATION OF FREQUENCY CURVES

A complete risk-based analysis sometimes requires an extension of flood-frequency curves beyond the 50-, 100-, or 500-year return periods, which is usually considered as defining the outer limits of the ability to make credible frequency estimates. As discussed in Appendix D, this is generally unjustified from available systematic flood flow records and records of historical floods. This poses a difficult problem if a frequency curve out to the PMF is to be employed unless some means is adopted to assign return periods to events such as the calculated PMF. Procedures for estimating the return period of PMF events are also discussed in Appendix D. Such direct methods are very involved. Lacking a simple alternative, a reasonable result can be obtained by considering the PMF to be the one in a million (10^6) year event, or the one in 10,000 (10^4) year event. Since use of the latter value would lead to higher estimates of risk-costs, it may be considered the more conservative assump-

242

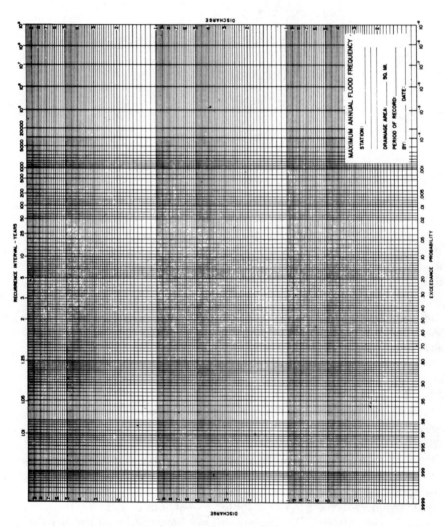

FIGURE E-1 Log normal plotting grid.

tion. Buehler (1984) recommended use of the 10^6 value. With such an assumption, one can extend the standard (Bulletin 17) frequency curve (on lognormal probability paper) to pass through the PMF with the specified exceedance probability. Figure E-1 provides an example of lognormal paper that can be used for this purpose. Here 10^4 and 10^6 represent a very conservative and a more reasonable value of the return period of PMF estimates, even though higher return period values have been suggested and in some cases are more likely (Newton, 1983). Ideally, these target values would be a function of the region of the country in which one was working, and the size of the river basin of concern for both of these factors should explain variations in the return periods of PMF estimates.

There are several reasonable pathways of extending the empirical frequency curve from the 100-year flood to the PMF. A simple linear extension of the empirical frequency curve from the 100-year flood to the PMF on lognormal paper is one reasonable procedure. The Bureau of Reclamation (1981a) uses a log-flow versus log-exceedance probability graph to extend the frequency curve; their guidelines also specify that the frequency curve should pass through a "box" bounded vertically by 40 and 60 percent of the PMF and horizontally by vertical lines drawn at the 200- and 500-year return periods. In practice, curves often are drawn to just pass through the lower right-hand corner of that box. Thus, the general effect of their procedure is to make the 0.4 PMF flood the 500-year event, irrespective of the size of the 100-year flood or the return period assigned to the PMF.

The Bureau's box constraint adds little to the analysis. While they may describe hydrologic experience in parts of southern California, these bounds should not be arbitrarily imposed upon flood-frequency curves for river basins in that region or elsewhere. Research should address the impact and advisability of using different axis scales to extend frequency curves to the PMF. However, because one is interpolating between the PMF with a specified exceedance probability and a specified 100-year flood, the differences should not be too great and should generally result in smaller variations in damage estimates than do use of either 10^{-4} or 10^{-6} as alternative values for the exceedance probability of the PMF.

Other problems arise when one attempts to make the flood-frequency curve approach the PMF flood asymptotically, as the Bureau suggests. Such efforts seem to reflect the mistaken belief that the PMF estimate is in fact the maximum possible flood that can occur. This is certainly not true (see discussion in Chapter 2 and Newton (1983)). Furthermore, the practice of treating the PMF estimates as the maximum possible flood can yield physically implausible bimodal flood distributions whose use can distort risk-cost analyses. This procedure implicitly assigns the values of 10^{-4} or 10^{-6} to the event that the PMF estimate occurs rather than to the event that the PMF is exceeded. Finally, this practice tempts one to think that if a dam were just designed for the PMF estimate, then there would be no risk of the dam overtopping. This is not the case. The PMF estimate is indeed a very large flood, but it can be exceeded.

Clearly, care should be exercised when extending flood-frequency relationships to PMF values. Additional research is clearly needed in this area. At present, reasonable and realistic risk investigations can be conducted by a linear extension (on lognormal paper or some reasonable alternative) of the frequency curve out through the PMF estimate, which is assigned a return period of 10^6 years, or the smaller and more conservative value of 10^4 years. By using alternative flood-frequency relationships that jointly span the reasonable range one can capture the sensitivity of the expected damage costs and, in particular, the ranking of the alternative projects, to our uncertainty as to the true flood-frequency relationship.

One should also remember that available flood records yield 100-year flood estimates with considerable uncertainty (Kite, 1977); thus it can be appropriate to include the sensitivity analysis, both alternative values of the exceedance probability of the PMF and also the 100-year flood. Procedures for calculating confidence intervals for the 100-year flood are available (Stedinger, 1983c).

EVALUATING RISKS

A risk-based analysis needs to consider the consequences and costs of reservoir operation (including damages from high lake levels and discharge, and also damage to the dam and from interruption of services) and the relative likelihood of such events. In general, four metrics are used to describe the consequences for each alternative considered:

1. likely loss of life;
2. economic damages from lake levels, releases, and damage to the dam;
3. the cost of actions associated with each modification of the dam, reservoir, and associated channels and any flood warning system; and
4. the cost of discontinued or interruptions in service due to damage to or the failure of the dam because of an extraordinary hydrologic event.

An American Society of Civil Engineers (ASCE) task-force (ASCE, 1973) has recommended that an economic value be assigned to loss of life so that all quantities can be added on an annual basis to obtain a total-cost function. While such a step is attractive and can be done as an extra computation, in general, initially displaying expected loss of life in a separate category is advisable.

The Bureau of Reclamation (1981a, 1984) provides a discussion of how risk-cost analysis can be performed. The key step is to provide a numerical approximation to the expected loss of life and the expected damage costs.

Comparison of project benefits in terms of only expected damages and loss of life is a problem in that it averages or integrates over many possible events. In addition and before calculating expected costs, engineers should compute and examine matrices such as that in Figure E-2, which can show expected loss of life and economic damages associated with each alternative project over a range of inflow floods. As noted by Interagency Committee on Dam Safety (ICODS) (1983, pp. 30-31), such calculations are necessary for risk-cost analyses.

One big unknown in any analysis of damages anticipated from dam overtopping and failure is the rate of breach development and the size of breach that will develop. Fortunately, we do not have many case histories of breach of large dams on which to base such estimates. A sensitivity analysis assuming a range of rate of development and size of breaches could show if these assumptions are critical for a specific case.

Intentionally breaching or destroying a dam, or widening or lowering a spillway, can result in increased discharges for modest 50- to 200-year return period floods, thereby increasing downstream damage cost associated with relatively likely events even though such structural modifications allow passage of the relatively improbable PMF without structural failure. On the other hand, given the size of near-PMF floods even without structural failure of the dam, sudden or progressive dam breach during a flood may have a

Flow Rate Proposal	10,000 (100-yr flood)	20,000	30,000	50,000	70,000	90,000 cfs (PMF)
Do nothing	D_{11} (L_{11})	D_{12} (L_{12})	D_{13} (L_{13})	D_{14} (L_{14})	D_{15} (L_{15})	D_{16} (L_{16})
Modify spillway—1						
Modify spillway—2						
Change operating policy plus modification—1						
Breach dam	D_{51} (L_{51})	D_{52} (L_{52})	D_{53} (L_{53})	D_{54} (L_{54})	D_{55} (L_{55})	D_{56} (L_{56})

FIGURE E-2 Sample decision matrix for risk analysis (expected damages D_{ij} associated with proposal i and inflow flood level j and also the corresponding expected loss of life).

very modest incremental impact on potential loss of life and economic damages. All these issues should be considered when selecting an appropriate modification of a structure and can be considered even before probabilities are employed to calculate expected loss of life and damage costs.

AN EXAMPLE

To illustrate the use of risk analyses and the balancing of the increment of property and lives at risk due to spillway inadequacies against retrofitting costs, we consider an example that is a generalization of several studies conducted by the Bureau of Reclamation. In this example, a reservoir constructed in the 1920s has a storage capacity 20,000 acre-feet. It has been found to have an inadequate spillway due to defects in the design of the spillway as well as a design discharge, which is substantially smaller than current PMF estimate. The PMF for the dam is currently estimated to be 120,000 cubic feet per second (cfs), while the 100-year flood is 20,000 cfs. The current spillway, given the design defects, is estimated to be capable of carrying 50,000 cfs. The minimum damage flood is 10,000 cfs, which is about the 50-year flood.

Alternative designs have been considered. One can (1) do nothing, (2) modify the existing spillway to pass 75,000 cfs, (3) rebuild the spillway and raise the dam so that it can pass the PMF, or (4) lower the spillway crest and lengthen it so that the full PMF can be passed without raising the dam. Table E-1 summarizes these options and their amortized annual costs.

The question is whether to stay with the current spillway with capacity of 50,000 cfs, to increase the capacity to 75,000 cfs for an annual cost of

TABLE E-1   Design Options and Costs for Illustrative Example

Option	Design Flow (cfs)	Annual Cost ($/yr)
1. Do nothing	50,000	0
2. Modify existing spillway	75,000	80,000
3. Rebuild spillway and raise dam	120,000	200,000
4. Rebuild spillway and lower or lengthen crest	120,000	120,000

$80,000, or to increase the capacity to the full PMF of 120,000 cfs at an annual cost of either $120,000 or $200,000.

A key ingredient in any decision may be the expected downstream (and sometimes upstream) property damages that result from large flood flows and reservoir operations. In this example, damages in the valley below the dam (designated by M) are assumed to rise monotonically toward an upper bound or maximum M if the dam does not fail. Any actual failure of this large dam is assumed to result in a flood wave that causes the maximum damages. The mathematical form of the damage function and other details of this analysis are provided at the end of Appendix D.

Clearly some trade-off is necessary between lives at risk, the possible cost of the project's destruction and the resulting loss of benefits, and the incremental flood damages that would result should the dam fail.

In this example, the incremental risk of loss of life due to dam failure during an extraordinary flood event is assumed to be negligible. Downstream residents would have been evacuated already or could be evacuated before the flood wave from a dam failure would reach them.

One must also consider the possible cost of replacing the dam should it fail and the cost of loss of services until the dam is again operational if that is the chosen course of action. The cost of reconstructing the dam and the present value of loss of services until the dam could be replaced are denoted by the letter L. Thus, if a flood overtops the reservoir, causing a breach, the total losses M + L are those due to downstream property damage and loss of the reservoir and the services not supplied until the reservoir can be rebuilt.

The matrices of costs that result from five different but reasonable combinations of M, L, and a third (damage-shape) parameter are shown in Table E-2. Each matrix represents the damage costs that occur from the flood flows or M + L if the dam is overtopped and by assumption fails. One must be careful when interpreting the cost functions because they are discontinuous at the spillway design flood. For example in case 1, if option 2 were pursued and a 75,000 cfs flood occurred, then downstream damages would be $9.6 million. However, a 76,000-cfs flood is assumed to overtop the reservoir causing total damages of $40 million.

In the first three cases in Table E-2 the committee has considered a range of L values representing a relatively low-cost dam with small loss of service costs to either a high-cost dam and/or high loss of service costs. Cases 4 and 5 correspond to low loss of service costs with two high downstream damage cost cases. Two things stand out in matrices: (1) the higher the spillway design discharge, the larger the range over which dam failure is avoided and damages stay below the maximum, and (2) the fourth inexpensive option of lowering, or lengthening, the spillway results in larger damage costs. What is not simple is the trade-off of annual modification costs in the $80,000–

TABLE E-2 Matrices of Damages (in millions) for Different Flows and Design Options

Design Option	Peak Flood Flow				
	20K	50K	75K	100K	120K
Case 1:	*M = $20 million, L = $20 million*				
1	1.9	6.6	40	40	40
2	1.9	6.6	9.6	40	40
3	1.9	6.6	9.6	12	13
4	3.6	11	15	17	18
Case 2:	*M = $20 million, L = $100 million*				
1	1.9	6.6	120	120	120
2	1.9	6.6	9.6	120	120
3	1.9	6.6	9.6	12	13
4	3.6	11	15	17	18
Case 3:	*M = $20 million, L = $400 million*				
1	1.9	6.6	420	420	420
2	1.9	6.6	9.6	420	420
3	1.9	6.6	9.6	12	13
4	3.6	11	15	17	18
Case 4:	*M = $100 million, L = $20 million*				
1	9.5	33	120	120	120
2	9.5	33	48	120	120
3	9.5	33	48	59	67
4	18	55	73	83	89
Case 5:	*M = $200 million, L = $20 million*				
1	5.9	23	220	220	220
2	5.9	23	35	220	220
3	5.9	23	35	47	56
4	9.8	36	55	72	85

NOTE: In the first four cases, the damage function shape parameter s (see Appendix D) equaled 1×10^{-5} for design options 1-3 and 2×10^{-5} for design option 4 (lower spillway crest); for case 5 the values were 3×10^{-6} and 5×10^{-6}, respectively.

$200,000 range with possible but very unlikely damages in the $10,000,000–$420,000,000 range.

To trade off dam failure prevention costs with possible flood damage and dam failure costs, an extended flood-frequency curve was constructed. A shifted exponential distribution was used to describe the flood-frequency distribution. Its two parameters were chosen so that 20,000 cfs was the 100-year flood (or 1 percent chance event) and the PMF of 120,000 cfs was the T-year event for T $= 10^4$ or 10^6. The expected damages due to flood events between the minimum damage flood of 10,000 cfs and the PMF were then calculated analytically (see Appendix D). The results are given in Table E-3 along with the sum of the expected damages plus the annualized construction costs from Table E-1.

Examination of Table E-3 illustrates a range of situations that might be encountered. In case 1, on an economic basis, no construction seems justified; even when the PMF is only the 10,000-year event, the least total cost option is to do nothing. This is consistent with the Table E-2 matrix for this case, which shows that damages in the event of a dam failure are of the same order of magnitude as damages without a failure.

Going to case 2, L increases from $20 million to $100 million. As a result, on an economic basis, option 2 with a 75,000-cfs spillway gives lowest indicated average annual cost if the PMF is only the 10,000-year flood, whereas to do nothing is indicated if the PMF is the million-year event.

Finally, in case 3 the cost of loss of the dam is very large, $400 million, and the inexpensive 120,000-cfs spillway design (option 4 involving lowering the crest) gives indicted lowest cost for T $= 10^4$, while the 75,000-cfs spillway design (option 2) shows lowest cost if T $= 10^6$. Unfortunately, the least cost solution in both cases 2 and 3 is sensitive to T.

Cases 4 and 5 consider situations where downstream damages are the important factor rather than the loss of the dam and the services it provides. In case 4, building the 75,000-cfs spillway (option 2) only shows lowest indicated cost if T is as small as 10^4. In case 5, with M $=$ $200 million, construction seems to give lowest indicated cost options, though different T values again lead to different decisions. However, in both cases 4 and 5, the lower crest 120,000-cfs spillway (option 4) is very unattractive because of the larger downstream damages it causes at all inflow levels up to and including the PMF. This stands in contrast to case 3, which had relatively small downstream damage potential, making the low-cost, lower crest spillway more attractive on an indicated cost basis. One should also remember that this analysis is based solely on the dollar values of property damage and excludes loss of life, which is assumed not to be a concern in this case.

USE OF EXPECTED COSTS

Actually calculating the expected costs and benefits associated with different designs provides some guidance or confirmation as to which design alternatives on balance are most attractive. However, an expected cost analysis does have some practical and philosophical problems.

On the practical side, at the extreme, one is multiplying estimates of large costs times rather poor estimates of small probabilities. The resultant prod-

TABLE E-3 Expected Damages for Different Design Options, Cases, and Return Periods

| | T Assigned to PMF | | | |
| | Expected Annual Damages ($1,000/yr) | | Expected Annual Damages Plus Construction Costs ($1,000/yr) | |
Option	$T = 10^4$	$T = 10^6$	$T = 10^4$	$T = 10^6$
Case 1:	*M = $20 million, L = $20 million*			
1	130	70	130	70
2	75	50	155	130
3	55	50	255	250
4	94	90	214	210
Case 2:	*M = $20 million, L = $100 million*			
1	320	120	320	120
2	130	55	210	135
3	55	50	255	250
4	95	90	215	210
Case 3:	*M = $20 million, L = $400 million*			
1	1050	310	1050	310
2	340	75	420	155
3	55	50	255	250
4	95	90	215	210
Case 4:	*M = $100 million, L = $20 million*			
1	460	300	460	300
2	320	250	400	330
3	275	245	475	445
4	470	450	590	570
Case 5:	*M = $200 million, L = $20 million*			
1	640	280	640	280
2	310	170	390	250
3	185	160	385	360
4	300	260	420	380

NOTE: Minimum cost value is underlined for each case and value of T.

uct may be a very imprecise estimate of the desired expected cost associated with extreme events. Thus, these calculations only provide a credible guide for decision making if sensitivity analyses considering alternative but credible costs and flood-frequency relationships yield a similar ranking of the alternatives.

An even more philosophical question is whether expected loss of life and expected damages are appropriate metrics to describe possible catastrophic events that are unlikely to occur in contrast to construction costs, which definitely will be experienced. Few argue against the use of probabilities in the analysis of recurring events without catastrophic consequences; examples are the analysis of the cost and benefits associated with frequent floods, year-to-year hydropower operations, or recurring navigational issues on waterways. However, the failure of a major dam in an extraordinary flood should not be a recurring event for which one can weigh the average annual costs that will be experienced from failures with the costs paid to reduce the likely severity and frequency of those failures. The issue is how much members of society are willing to pay to avoid such unlikely events. It is highly plausible that they are willing to pay more than the expected cost.

The cost society should bear to avoid the consequences of dam failure (both immediate and those due to the inability of the facility to continue to provide the anticipated services) is difficult to determine. When loss of life is a major issue, most people agree that little risk is acceptable and PMF standards for safety are appropriate. But why not a little more or a little less?

LOSS OF LIFE CALCULATIONS

The balancing of reservoir construction and retrofitting costs against the expected incremental property damage costs from improbable but possible dam failure is a balancing of monies expended for protection versus properties damaged for which monetary compensation can be provided. This comparison can be performed on an annual cost basis (amortized reservoir construction costs versus the expected incremental damages avoided per year) given an interest rate and a planning period. Alternatively, given the interest rate and a planning period, one can make the equivalent comparison between the present value of the two time streams. Loss of life considerations pose fundamentally different considerations, particularly in a time value context. First, one must be willing to specify the amount of federal monies that society is willing to expend to avoid an accidental death or on average one accidental death that could occur during the next fiscal year. While no one is anxious to specify a particular dollar value for such a death, reasonable ranges can be surmised.

The question dam safety raises is, how much is society willing to pay today to avoid an accidental death that could occur in 25, 50, or 100 years? If the interest is 10 percent per year (or 5 percent per year) and society would be willing to pay as much as $1 million per expected accidental death avoided in the immediate future, would society be willing to pay only the present value of $1 million in 50 years at 10 percent per year, which is $8,519 (or $87,200 for i = 5 percent), to avoid an expected death at that time? Certainly the first value is too small and probably also the second.

Discounting the value of life with the discount rate for federal expenditures does not seem appropriate. An alternative is not to discount the value assigned to loss of life or to discount it very little. If this is done, one must be careful not to compare amortized capital and construction costs with annual expected loss of life values; that would be equivalent to discounting the value of life at the discount rate for construction costs.

What are the consequences of not discounting loss of life? If loss of life is not discounted, then it is easiest to compare the cost of construction and spillway modification costs with the present value of incremental damages avoided (calculated with the interest rate for property) and the incremental expected number of accidental deaths avoided per year times the number of years in the planning period. Thus, if a dam modification in expectation saves 0.02 lives per year, this is equivalent to saving 1 life in 50 years, 2 lives in 100 years, and 4 lives in 200 years. Because the number of expected lives the dam modification is credited with saving increases linearly with time, and the dam may be around for a long time, the value of life can loom large in the analysis. Of course, if the expected lives lost per year is only 10^{-4}, then after 200 years the expected loss of life is still only 0.02.

With these considerations before us and given the long service lives of many dams, there is no rigorous method to decide how to properly balance construction and spillway modification costs with loss of life values. Clearly, absolute safety cannot in general be provided. However, our society seems willing to spend large sums of money so that federal dams and other large projects do not put people's lives at risk without their explicit consent, regardless of what people choose to do voluntarily. Clearly the federal government wants to continue to protect the lives of its citizens.

ADVANTAGES AND LIMITATIONS OF RISK ANALYSIS APPROACH

In Chapter 5 some of the limitations and advantages of the risk analysis approach are outlined. For example, a key feature of the matrix decision approach and expected cost analyses is that the selected design is not confined to a specified SEF. The analysis probably would include the PMF as one of the safety evaluation floods, but no single flood is specified as the unique inflow hydrograph against which a design should be tested. This is as it should be, because no single event can capture the character of all of the small and large floods to which a dam may be subjected. Strict reliance on PMF as inflow design floods or some other unique safety evaluation flood in a real sense can distract designer and public attention from the real flood risk, which is probably due to less extreme but much more likely hydrologic events as well as from structural and foundation problems with the dam. which can also cause dam failure. Kirby (1978) suggests that just "the concept of probable maximum flood diverts our attention toward the illusion of absolute safety and away from the hard necessity to accept risk and minimize it by balancing opposing risk." The matrix decision table may make these trade offs more apparent.

Dams are major social investments, and their failures due to the occurrence of large floods is of much concern both because of the loss of the investment embodied by the dam and the likely loss of life and property such a dam failure might cause. In the design against extreme hydrologic events, dam designers have had to assess the hydrologic risk present at a particular dam site. There are basically two approaches that can or have been applied to assess that risk. One is to consider the probable maximum flood event that would occur following a probable maximum precipitation event in conjunction with other unfavorable factors. The second approach would entail the probabilistic analysis of a wide range of scenarios so as to construct a flood-frequency relationship. The Harriman Dam analysis (see Appendix A, Yankee Atomic Electric Company) illustrates such an effort.

While there are substantial difficulties in estimating the return period of the PMF or in estimating the 10,000-year flood using probabilistic techniques, all of the difficulties are also associated with characterizing worst case analyses such as the probable maximum flood calculation. Both approaches require judgment. Neither are automatic. To be done satisfactorily, both approaches require understanding of the underlying processes generating extreme events.

Traditionally, the appropriate level of safety was often determined by professionals involved in the construction and regulatory agencies of the federal government. Based on their collective and joint consideration, reservoir designs were adopted that reflected reasonable standards of safety. However, decision making patterns in our society are changing, so that a large audience, including more of the general public, is becoming involved in important social decisions. There are two dangers. The first is that the decision making and safety standard setting process will come to include so many participants and interests that acceptable and appropriate decisions cannot be made. On the other hand, the decision making process will lose its social legitimacy if too few groups or only a select group can become involved. This can occur if professionals treat important social decisions as simply technical issues that they feel authorized to resolve. While society once may have delegated such authority to design engineers, social decision patterns have and are continuing to change. Thus, federal dam-building agencies need to remain open to and should encourage public discussion of their design practices to assure that their practices enjoy general public and professional support. One advantage of the matrix decision approach is that, in the context on a particular decision, it displays the range of alternatives that were considered and illustrates the performance of the dam over a wide range of floods.

Historically, dam designers have appropriately been concerned that their profession not be discredited nor public confidence lost by the failure of dams during large floods or earthquakes. However, in view of the greater technical expertise available today and society's current outlook on the allocation of resources, safety factors and overdesign practices that once may have been appropriate may no longer be reasonable. Thus, government agencies should carefully determine if the marginal increments in damages that result from changes in operating rules, structural modifications, and dam failures on balance indicate that structural modification and operating rule changes for existing dams are both justified and well advised. The procedures discussed in this Appendix can facilitate such determinations.

Glossary

Several terms used in this report are peculiar to or most often used in rather specialized technical disciplines. Others are used in special, restricted senses. Some such terms are defined when first used; others are not. For convenience of the reader a number of those technical terms are defined here.

DAM   A barrier built across a watercourse for impounding or diverting the flow of water.

DRAINAGE AREA   The area that drains naturally to a particular point on a river.

EMERGENCY ACTION PLAN   A predetermined plan of action to be taken to reduce the potential for property damage and loss of lives in an area affected by a dam break.

EPICENTER   That point on the earth's surface that is directly above the focus of an earthquake.

FAILURE   An incident resulting in the uncontrolled release of water from a dam.

FLOOD PLAIN   An area adjoining a body of water or natural stream that has been or may be covered by floodwater.

FLOOD ROUTING   The determination of the modifying or attenuating effect of passage of a flood through a valley, channel, or reservoir.

FOCUS (HYPOCENTER)   The point within the earth that is the center of an earthquake and the origin of its elastic waves.

FREEBOARD   The vertical distance between a stated water level and the top of a dam.

255

HAZARD   A source of danger. In other words, something that has the potential for creating adverse consequences.

HYDROGRAPH   A graphical representation of discharge, stage, or other hydraulic property with respect to time for a particular point on a stream. (At times the term is applied to the phenomenon the graphical representation describes; hence, a flood hydrograph is the passage of a flood discharge past the observation point.)

INTENSITY SCALE   An arbitrary scale to describe the degree of shaking at a particular place. The scale is not based on measurement but on a descriptive scale by an experienced observer. Several scales are used (e.g., the Modified Mercalli scale, the MSK scale), all with grades indicated by Roman numerals from I to XII.

LOADING CONDITIONS   Events to which the dam is exposed, e.g., earthquake, flood, gravity loading.

MAGNITUDE (see also RICHTER SCALE)   A rating of the size of an earthquake by numerical values, such as $M5.6$, $M8.2$, etc. The magnitude number is calculated by means of the logarithm of the amplitude of matrices recorded by a standard seismograph at a known distance from the origin of the earthquake. Each higher whole number expresses an amount of energy released that is approximately 60 times larger than that expressed by the preceding whole number, for example an $M6$ earthquake will release about 60 times the energy of an $M5$ earthquake.

MAXIMUM CREDIBLE EARTHQUAKE (MCE)   The severest earthquake that is believed to be possible at the site on the basis of geologic and seismological evidence. It is determined by regional and local studies that include a complete review of all historic earthquake data of events sufficiently nearby to influence the project, all faults in the area, and attenuations from causative faults to the site.

ONE-HUNDRED-YEAR (100-YEAR) EXCEEDANCE INTERVAL FLOOD   The flood magnitude expected to be equaled or exceeded on the average of once in 100 years. It may also be expressed as an exceedance frequency with a 1 percent chance of being exceeded in any given year.

OPERATING BASIS EARTHQUAKE   More moderate than the MCE and may be selected on a probabilistic basis from regional and local geology and seismology studies as being likely to occur during the life of the project. Generally, it is at least as large as earthquakes that have occurred in the seismotectonic province in which the site is located.

PROBABILITY   The likelihood of an event's occurring.

PROBABLE MAXIMUM FLOOD (PMF)   The flood that may be expected from the most severe combination of critical meteorologic and hydrologic conditions that are reasonably possible in the region. This term as used in official documents of the Corps of Engineers identifies estimates of hypothetical flood characteristics (peak discharge, volume, and hydrograph shape) that are considered to be the most severe "reasonably possible" at a particular location, based on relatively comprehensive hydrometeorological analyses of critical runoff-producing precipitation (and snowmelt, if pertinent) and hydrologic factors favorable for maximum flood runoff. (ONE-HALF PMF   That flood with a peak flow equal to one-half of the peak flow of a probable maximum flood.)

PROBABLE MAXIMUM PRECIPITATION (PMP)   Theoretically, the greatest depth of precipitation for a given duration that is physically possible over a given size storm area at a particular geographical location at a certain time of the year. (This definition is a 1982 revision and results from mutual agreement among the National Weather Service, the U.S. Army Corps of Engineers, and the Bureau of Reclamation.)

PSEUDO-ADIABATIC   A term applied to a vertical temperature variation in the atmosphere in which the temperature at any elevation is that which would be attained if a unit mass of air at ground surface were carried aloft to that elevation and allowed to expand to ambient pressure without loss of heat.

RESERVOIR ROUTING   The computation by which the interrelated effects of the inflow hydrography, reservoir storage and discharge from the reservoir are evaluated.

RICHTER SCALE   A scale proposed by C. F. Richter to describe the magnitude of and earthquake by measurements made in well-defined conditions and with a given type of seismograph. The zero of the scale is fixed arbitrarily to fit the smallest recorded earthquakes. The largest recorded earthquake magnitudes are near 8.7 and are the result of observations and not an arbitrary upper limit like that of the intensity scale.

RISK   The likelihood of adverse consequences.

RISK ASSESSMENT   As applied to dam safety, the process of identifying the likelihood and consequences of dam failure to provide the basis for informed decisions on a course of action.

RISK COST (EXPECTED COST OF FAILURE)   The product of the risk and the monetary consequences of failure.

SAFETY EVALUATION EARTHQUAKE (SEE)   The earthquake, expressed in terms of magnitude and closest distance from the dam site, or in terms of the characteristics of the time history of free-field ground motions, for which the safety of the dam and critical structures associated with the dam is to be evaluated. In many cases this earthquake will be the maximum credible earthquake to which the dam will be exposed. However, in other cases, where the possible sources of ground motion are not readily apparent, it may be a motion with prescribed characteristics selected on the basis of a probabilistic assessment of the ground motions that may occur in the vicinity of the dam. It should be demonstrated that the dam can withstand this level of earthquake shaking without release of water from the reservoir.

SAFETY EVALUATION FLOOD (SEF)   The largest reasonable hypothetical water inflow for which the safety of a dam and appurtenant structures is to be evaluated; it should be demonstrated that this flood level can be accommodated through storage, spillway releases, releases through other outlet works or limited and acceptable overtopping without causing failures of the structure or release of impounded water.

SEISMIC INTENSITY   See intensity scale.

SPILLWAY   A structure over or through which flood flows are discharged. If the flow is controlled by gates, it is considered a controlled spillway; if the elevation of the spillway crest is the only control, it is considered an uncontrolled spillway.

SPILLWAY DESIGN FLOOD (SDF)   The largest flood that a given project is designed to pass safely.

STORAGE   The retention of water or delay of runoff either by planned operation, as in a reservoir, or by temporary filling of overflow areas, as in the progression of a flood crest through a natural stream channel.

STORAGE RESERVOIR   A reservoir that is operated with changing water level for the purpose of storing and releasing water.

UNIT HYDROGRAPH   The hydrograph of flow from a watershed produced by a unit volume of runoff generated during a specific period of rainfall excess in the watershed. In the United States a unit hydrograph normally represents 1 inch of runoff volume. Hence, a 6-hour unit hydrograph represents runoff from the watershed resulting from rainfall excess (rainfall minus infiltration and other losses) of 1 inch over a 6-hour period.

APPENDIX G

References and Bibliography

Ahlberg, J. H., E. N. Nilson, and J. L. Walsh (1967). The Theory of Splines and Their Application. New York: Academic Press.

Algermissen, S. T. (1969). Seismic risk studies in the United States. Paper presented at 4th World Conference on Earthquake Engineering, Santiago, Chile.

Algermissen, S. T. (1983). An Introduction to the Seismicity of the United States. EERI Monograph. Berkeley, California: Earthquake Engineering Research Institute. 148 pp.

Algermissen, S. T. and D. M. Perkins (1976). A Probabilistic Estimate of Maximum Acceleration in Rock in the Contiguous United States. Open-File Report 76-416. Reston, Virginia: U.S. Geological Survey, U.S. Department of the Interior. 45 pp.

Algermissen, S. T., D. M. Perkins, P. C. Thenhaus, S. L. Hanson, and B. L. Bender (1982). Probabilistic Estimates of Maximum Acceleration and Velocity in Rock in the Contiguous United States. Open File Report 82-1033. Reston, Virginia: U.S. Geological Survey, U.S. Department of the Interior.

American Meteorological Society (1959). Glossary of Meteorology, Ralph Huschke, ed. Boston, Massachusetts.

American Society of Civil Engineers (1973). Re-evaluating Spillway Adequacy of Existing Dams. New York: Task Committee of the Committee on Hydrometeorology, Hydraulics Division.

Benjamin, J. R. and C. A. Cornell (1970). Probability, Statistics and Decisions for Civil Engineers. New York: McGraw-Hill.

Benson, M. A. (1968). Uniform flood-frequency estimating methods for federal agencies. Water Resources Research 4(5):891–908.

Buehler, B. (1984). Discussion on paper by Donald W. Newton, "Realistic assessment of maximum flood potentials." Journal of Hydraulic Engineering 110(8):1166–1168.

Bureau of Reclamation (1981a). Criteria for Selecting and Accommodating Inflow Design Floods for Storage Dams. ACER Technical Memo 1. Washington, D.C.: U.S. Department of the Interior. 43 pp.

Bureau of Reclamation (1981b). Freeboard Criteria and Guidelines for Computing Freeboard

259

Allowances for Storage Dams. ACER Technical Memo 2. Washington, D.C.: U.S. Department of the Interior. 70 pp.

Bureau of Reclamation (1984). Design Standards, No. 13, Chapter 14. Washington, D.C.: U.S. Department of the Interior.

Cohn, T. and J. R. Stedinger (1983). The use of historical flood records in flood frequency analysis. Paper presented at American Geophysical Union Meeting, San Francisco, California, December 1983.

Committee on Safety of Existing Dams (1983). Safety of Existing Dams—Evaluation and Improvement. Prepared under auspices of Water Science and Technology Board, Commission on Engineering and Technical Systems, National Research Council. Washington, D.C.: National Academy Press. 374 pp.

Condie, R. and K. A. Lee (1982). Flood Frequency Analysis with Historic Information, Journal of Hydrology 58:47–61.

Covello, V. T., W. G. Flamm, J. V. Rodricks, and R. G. Tardiff (eds.) (1983). The Analysis of Actual Versus Perceived Risks. New York: Plenum.

Dower, R. C. (1983). Survey of Federal Agency Research and Application of Economic Benefits Analysis. Prepared by the Environmental Law Institute for the U.S. Environmental Protection Agency.

Fenn, D. D. (in press). Probable Maximum Precipitation Estimates for Johns Creek Basin above Dewey Dam. NOAA Technical Memorandum NWS HYDRO 41. Silver Spring, Maryland: National Weather Service, National Oceanic and Atmospheric Administration, U.S. Department of Commerce.

Fuller, W. E. (1914). 1914 Flood flows. Transaction of the American Society of Civil Engineers ASCE 77:564–617.

Gillette, R. (1974). Nuclear safety: Calculating the odds of disaster. Science 185:838–839.

Gould, B. W. (1973). Discussion of "Bias in computed flood risk." Journal of Hydraulics Division, ASCE 99(HY1):270–273.

Hansen, E. M., J. T. Riedel, and F. K. Schwarz (1977). Probable Maximum Precipitation Estimates—Colorado River and Great Basin Drainages. Hydrometeorological Report 49. Silver Spring, Maryland: National Weather Service, National Oceanic and Atmospheric Administration, U.S. Department of Commerce. 161 pp.

Hansen, E. M., L. C. Schreiner, and J. F. Miller (1982). Application of Probable Maximum Precipitation Estimates—United States East of the 105th Meridian. Hydrometeorological Report 52. Washington, D.C.: National Weather Service, National Oceanic and Atmospheric Administration, U.S. Department of Commerce. 168 pp.

Hershfield, D. M. (1961). Estimating the probable maximum precipitation. Proceedings, ASCE, Journal of Hydraulics Division 87:99–106.

Hershfield, D.M. (1965). Methods of estimating probable maximum precipitation. Journal of American Waterworks Association 57:965–972.

Ho, F. P. and J. T. Riedel (1980). Seasonal Variation of 10-Square Mile Probable Maximum Precipitation Estimates—United States East of the 105th Meridian. Hydrometeorological Report 53. Washington, D.C.: National Weather Service. 89 pp.

Hornbeck, R. W. (1975). Numerical Methods. Englewood Cliffs, New Jersey: Prentice-Hall.

Housner, G. W. and P. C. Jennings (1982). Engineering Monographs on Earthquake Criteria, Structural Design, and Strong Motion Records, Vol. 4, Earthquake Design Criteria, Berkeley, California: Earthquake Engineering Research Institute.

Huber, P. (1983). The old-new division in risk regulation. Virginia Law Review 69:1025–1107.

Interagency Advisory Committee on Water Data (1982). Guidelines for Determining Flood Flow Frequency. Washington, D.C.

Interagency Committee on Dam Safety (1983). Proposed Federal Guidelines for Selecting and

Accommodating Inflow Design Floods for Dams. Draft prepared by Working Group on Inflow Design Floods, Subcommittee 1 of ICODS.

International Commission on Large Dams (1974). Earthquake Committee.

International Commission on Large Dams (1975). A Review of Earthquake Resistant Design of Dams. Bulletin No. 27. U.S. Committee on Large Dams, Boston, Massachusetts: Chas. T. Main, Inc.

International Conference of Building Officials (1979). Uniform Building Code, 1979 Edition. Whittier, California.

Jarrett, R. D. and J. E. Costa (1982). Multidisciplinary Approach to the Flood Hydrology of Foothill Streams in Colorado. Report on International Symposium on Hydrometeorology (June 1982). Denver, Colorado: American Water Resource Association, pp. 565–560.

Jennings, A. H. (1952). Maximum 24-Hour Precipitation in the United States. Technical Paper 16. Washington, D.C.: U.S. Weather Bureau, U.S. Department of Commerce. 284 pp.

Kirby, W. (1978). Predictions of streamflow hazards. Pp. 202–215 in Geophysical Predictions. Washington, D.C.: National Academy Press.

Kite, G. W. (1977). Frequency and Risk Analyses in Hydrology. Fort Collins, Colorado: Water Resources Publications.

Lave, L. B. (1981). The Strategy of Social Regulation: Decision Frameworks for Policy. Washington, D.C.: Brookings Institution.

Miller, J. F. (1963). Probable Maximum Precipitation and Rainfall-Frequency Data for Alaska. Technical Paper 47. Washington, D.C.: U.S. Weather Bureau, U.S. Department of Commerce. 69 pp.

Miller, J. F., R. H. Frederick, and R. J. Tracey (1973). Precipitation-frequency atlas of the western United States. NOAA Atlas 2. Silver Spring, Maryland: National Weather Service, National Oceanic and Atmospheric Administration, U.S. Department of Commerce. 11 vols.

Miller, J. F., E. M. Hansen, and D. D. Fenn (1984). Probable Maximum Precipitation for the Upper Deerfield River Drainage Massachusetts and Vermont. NOAA Technical Memorandum NSW HYDRO 39. Silver Spring, Maryland: National Weather Service, National Oceanic and Atmospheric Administration, U.S. Department of Commerce. 36 pp.

Myers, V. A. (1967). The Estimation of Extreme Precipitation as the Basis for Design Floods—Resume of Practice in the United States. Pp. 84–101 in Publication 84, Symposium at Leningrad (1967). Gentbrugge, Belgium: International Association of Scientific Hydrology.

New Columbia Encyclopedia (1975). Fourth ed., William H. Harris and Judith S. Levey, eds. New York: Columbia University Press.

Newton, D. W. (1983). Realistic assessment of maximum flood potentials. ASCE Journal of Hydraulic Engineering 109(6):905–918.

Riedel, J. T. (1977). Assessing the probable maximum flood. Water Power and Dam Construction (December).

Riedel, J. T. and L. C. Schreiner (1980). Comparison of Generalized Estimates of Probable Maximum Precipitation with Greatest Observed Rainfalls. NOAA Technical Report NWS 25. Silver Spring, Maryland: National Weather Service, National Oceanic and Atmospheric Administration, U.S. Department of Commerce. 66 pp.

Riedel, J. T., J. F. Appleby, and R. W. Schloemer (1956). Seasonal Variation of the Probable Maximum Precipitation East of the 105th Meridian for Areas of 10 to 1,000 Square Miles and Durations of 6, 12, 24, and 48 Hours. Hydrometeorological Report 33. Washington, D.C.: U.S. Weather Bureau, U.S. Department of Commerce. 58 pp.

Schreiner, L. C. and J. T. Riedel (1978). Probable Maximum Precipitation Estimates—United States East of the 105th Meridian. Hydrometeorological Report 51. Silver Spring, Maryland: National Weather Service, National Oceanic and Atmospheric Administration, U.S. Department of Commerce. 87 pp.

Schwarz, F. K. (1963). Probable Maximum Precipitation in the Hawaiian Islands. Hydrometeorological Report 39. Washington, D.C.: Weather Bureau, U.S. Department of Commerce. 98 pp.

Seed, H. B. and I. M. Idriss (1982). Ground Motions and Soil Liquefactions During Earthquakes. Berkeley, California: Earthquake Engineering Research Institute. 134 pp.

Stedinger, J. R. (1983a). Design events with a specified flood risk. Water Resources Research 19(April): 511–522.

Stedinger, J. R. (1983b). Estimating a regional flood frequency distribution. Water Resources Research 19(April): 503–510.

Stedinger, J. R. (1983c). Confidence intervals for design events. Journal of Hydraulics Division, ASCE 109(HY1): 13–27.

Thaler, R. and W. Gould (1982). Public Policies Toward Lifesaving: Should Consumer Preferences Rule? Journal of Policy Analysis and Management 1(2):223–242.

Tschantz, B. A. (1983). Report on Review of State Non-Federal Dam Safety Programs. Washington, D.C.: Federal Emergency Management Agency.

Tschantz, B. A. (1984). Updated Review Summary of State Non-Federal Dam Safety Programs. Washington, D.C.: Federal Emergency Management Agency.

U.S. Army Corps of Engineers (1945). Storm Rainfall in the United States. Washington, D.C.

U.S. Army Corps of Engineers (1982). Report of the Chief of Engineers to the Secretary of the Army on the National Program of Inspection of Non-Federal Dams. Washington, D.C.

U.S. Atomic Energy Commission (1974). An Assessment of Accident Risks in U.S. Commercial Nuclear Power Plants. Washington, D.C.

U.S. Committee on Large Dams (1970). Criteria and Practices Utilized in Determining the Required Capacity of Spillways. Report of USCOLD Committee on Failures and Accidents to Large Dams other than in Connection with the Foundations.

U.S. Department of the Interior (1982). Guidelines for Determining Flood Flow Frequency. Reston, Virginia: U.S. Geological Survey, Office of Water Data Coordination.

U.S. National Weather Service (1943–1984). National Oceanic and Atmospheric Administration Hydrometeorological Reports 1–55. (A series of PMP Studies for various drainages of regions. Many were not published, and many are out of print. Some are listed by authors elsewhere in this bibliography.) Washington, D.C.

U.S. National Weather Service (1968). Climatic Atlas of the United States. Washington, D.C.: U.S. Department of Commerce.

U.S. Nuclear Regulatory Commission (1981). Proceedings of Workshop on a Proposed Safety Goal by the NRC, July 23–24, 1981. Washington, D.C.: U.S. Government Printing Office, Division of Technical Information and Document Control NUREG/CP-0020.

U.S. Weather Bureau (1947). Generalized Estimates of Maximum Possible Precipitation Over the United States East of the 105th Meridian. Hydrometeorological Report 23. Washington, D.C.: U.S. Department of Commerce. Pp. 5–6.

U.S. Weather Bureau (1960). Generalized Estimates of Probable Maximum Precipitation West of the 105th Meridian. Technical Report 38. Washington, D.C.: U.S. Department of Commerce.

U.S. Weather Bureau (1961a). Interim Report—Probable Maximum Precipitation in California. (Including revision of 1969.) Hydrometeorological Report 36. Washington, D.C.: U.S. Department of Commerce. 101 pp.

U.S. Weather Bureau (1961b). Generalized Estimates of Probable Maximum Precipitation and Rainfall-Frequency Data for Puerto Rico and Virgin Islands. Technical Paper 42. Washington, D.C.: U.S. Department of Commerce. 94 pp.

U.S. Weather Bureau (1966a). Probable Maximum Precipitation Estimates—Northwest States. (Including revision of 1981.) Hydrometeorological Report 43. Washington, D.C.: U.S. Department of Commerce, Environmental Science Services Administration. 118 pp.

U.S. Weather Bureau (1966b). Meteorological Conditions for the Probable Maximum Flood on the Ukon Above Rambort, Alaska. Hydrometeorological Report 42. Washington, D.C.: U.S. Department of Commerce.

Vogel, R. and J. R. Stedinger (1984). Floodplain Delineation in Ice Jam Prone Regions. Journal of Water Resources, Planning and Management 110(WR2):206–219.

Wall, I. B. (1974). Probabilistic assessment of flooding hazard for nuclear power plants. Nuclear Safety 15(4):399–408.

Wallis, J. R. (1980). Risk and uncertainties in the evaluation of flood events for the design of hydrologic structures. Paper presented at the Seminar on Extreme Hydrological Events: Floods and Droughts, Centro di Cult. Sci. "E. Majorana," Erice, Italy, March 22–25, 1980.

Westergaard, H. M. (1933). Water pressures on dams during earthquakes. Transactions ASCE 98:418–433.

World Meteorological Organization (1969). Manual for Depth-Area-Duration Analysis of Storm Precipitation. WMO Report 129. Geneva, Switzerland.

World Meteorological Organization (1973). Manual for Estimation of Probable Maximum Precipitation. Operational Hydrology Report 1. Geneva, Switzerland.

Zangar, C. N. (1952). Hydrodynamic Pressures on Dams Due to Earthquake Effects. Engineering Monograph 11. Washington, D.C.: Bureau of Reclamation.

Biographical Sketches of Committee Members

MEMBERS

KEIITI AKI received his B.S. in 1952 and his Ph.D. in geophysics in 1958. Until recently he was a professor of geophysics at the Massachusetts Institute of Technology; he is now at the University of Southern California, Department of Geological Sciences. From 1960 to 1962 he was research fellow of seismology, and from 1960 to 1965 he was associate professor of seismology, both at the Earthquake Research Institute at the University of Tokyo. Dr. Aki's research is in geophysical research directed toward predicting, preventing, and controlling earthquake hazard; seismic wave propagation; earthquake statistics; and other areas. Dr. Aki is a member of the National Academy of Sciences.

DONALD H. BABBITT received a B.S. in engineering from the University of California at Berkeley and has been employed by the State of California, Department of Water Resources since 1957. Mr. Babbitt's experience includes designing or supervising design of eight dam embankments including 390-foot-high Pyramid Dam and 120-foot-high, 11,600-foot-long Perris Dam, both in southern California. Since 1976 he has been in the Division of Safety of Dams, in charge of a section reviewing design of dams and modifications to dams and reevaluating existing dams, primarily for seismic stability and spillway capacity.

DENIS BINDER received his J.D. from the University of San Francisco Law School in 1970. He earned an LL.M. in 1971 and an S.J.D in 1973 from

264

the University of Michigan. He is presently professor of law at Western New England College and has done research and published considerably in the area of environmental law, particularly dam safety.

CATALINO B. CECILIO is a senior civil engineer in charge of the hydrologic engineering group in the Civil Engineering Department of the Pacific Gas and Electric Company in San Francisco, California. He holds a B.S. degree in civil engineering and is registered as a professional engineer in the state of California. His job responsibilities and work experience since 1969 with Pacific Gas and Electric, which owns some 200 dams, has been in the hydraulics and hydrology of dam safety, with principal expertise on design floods up to and including the probable maximum flood and dam-break analysis. His expertise includes floodplain evaluation and the impact on the hydrologic environment of plant construction. He is principal codeveloper of the ANS 2.8 *American National Standards for Determining Design Basis Flooding at Power Reactor Sites*, first issued in 1976 by the American National Standard Institute and revised in 1981.

ALLEN T. CHWANG received his Ph.D. in 1971 from the California Institute of Technology and is a registered professional engineer in California. His academic experience includes research associate in engineering science at the California Institute of Technology; associate professor, Institute of Hydraulic Research, University of Iowa, and presently, professor, Institute of Hydraulic Research at the University of Iowa, Iowa City. Some of Dr. Chwang's publications are *Hydrodynamic Pressures on Sloping Dams During Earthquakes, Part 1, Momentum Method, Hydrodynamic Pressure on an Accelerating Dam and Criteria for Cavitation*, and *The Effect of Stratification on Hydrodynamic Pressures on Dams*.

MERLIN D. COPEN is presently an international consultant on concrete dams. Mr. Copen is retired head of Concrete Dams Section, Bureau of Reclamation E&R Center, Denver. He received his B.S. (Utah State) and M.S. (Kansas State) in civil engineering. Mr. Copen has a worldwide reputation as a designer of arch dams. He has authored publications on subjects such as earthquake loading for design of thin arch dams.

LLEWELLYN L. CROSS is the chief hydrologist for Chas. T. Main, Inc., in charge of all hydrometeorological and related work. He has more than 30 years of experience in hydrologic and hydrometeorological studies and designs. Mr. Cross holds a B.S. in civil engineering from Tufts University and is a member of the U.S. Committee on Large Dams and of the USCOLD Committee on Hydraulics of Spillways. At Chas. T. Main, Inc., he is responsible for hydrologic studies integral to the determination of spillway design floods, diversion floods, and reservoir yield studies for hydroelectric projects in the United States and abroad.

CHARLES H. GARDNER received an M.S. in geology from Emory University in 1961. He has studied civil and mining engineering and is a registered professional engineer and a certified professional geologist. He was chief geologist for International Minerals and Chemicals, Florida Phosphate Operations, and the chief geologist for Law Engineering Testing Company in Atlanta and Raleigh, which included responsibility for dam design and inspection of projects. Since 1976 he has been chief of the Land Quality Section of the North Carolina Department of Natural Resources and has overall responsibility for the state's programs in dam safety, mining, and sedimentation. Mr. Gardner is responsible for an inspection program covering 3,700 dams and has reviewed more than 400 dam design and repair plans. Mr. Gardner is a member of the U.S. Committee of the International Commission on Large Dams.

GEORGE W. HOUSNER received his B.S. degree from the University of Michigan and his Ph.D. from the California Institute of Technology, where he has served on the staff since 1945. His specialty is in earthquake engineering research, and he is a consultant on earthquake resistant design of major engineering projects. He is a member of the National Academy of Engineering (since 1965) and of the National Academy of Sciences. Dr. Housner has been a member and chairman of many National Research Council committees studying problems relative to earthquake engineering and has served as president of both the Earthquake Engineering Research Institute and the International Association for Earthquake Engineering. Dr. Housner is the author of three textbooks and 105 technical papers.

LESTER B. LAVE received his Ph.D. in economics from Harvard University and has been with Carnegie-Mellon University since 1963. He is presently professor of economics and head of the Department of the Graduate School of Industrial Administration. Dr. Lave has held various consulting positions with the Rand Corp., Resources for the Future, and the Department of Defense. He has served as a member of several National Research Council committees and is a member of the Institute of Medicine.

WALTER R. LYNN received the BSCE degree from Miami (Florida) in 1950, an MSSE from North Carolina in 1954, and a Ph.D. from Northwestern in 1963. He has had broad academic and consulting experience in water resources, sanitary and systems engineering and is currently professor and director of the Program of Science, Technology and Society at Cornell University. He is presently chairman of the Water Science and Technology Board and also past chairman of the National Research Council Committee to Review the Washington Metropolitan Area Water Supply Study.

DOUGLAS E. MACLEAN received his B.A. from Stanford University and his Ph.D. in philosophy from Yale University in 1977. Dr. MacLean has delivered many papers and lectures on the subject of risk assessment and decision making since 1979. He was a lecturer at Yale University from 1973 to 1975; lecturer at Livingston College, Rutgers University, from 1975 to 1976; a research associate and then senior research associate from 1977 to present at the Center for Philosophy and Public Policy at the University of Maryland. He received an award in 1980 from the National Science Foundation for his research on "Risk and Consent."

OTTO W. NUTTLI received B.S., M.S., and Ph.D. (1953) degrees in geophysics from Saint Louis University. He has since held various positions at that university. He is presently professor of geophysics in the Earth and Atmospheric Sciences Department. He is expert in earthquake seismicity, particularly in eastern North America.

JOHN T. RIEDEL attended the College of Idaho, Santa Ana Junior College, and the University of California, graduating with a B.S. in meteorology in 1947. He served as a weather officer in the Army Air Corps, and then began working in the Hydrologic Services Division, U.S. Weather Bureau, in 1947. He continued in this office until retirement in 1980. His career emphasis in the Weather Service was with extreme meteorological criteria important to planning and designing water control structures.

GURMUKH S. SARKARIA received a BSCE from Punjab University, India (1945), an MCE from Brooklyn Polytechnic Institute (1947), and an M.S. from Harvard (1948). He was with Bureau of Reclamation for 4 years and the Bhakra Dam, India, design group for 4 years. With International Engineering Company since 1956, he has served as vice president for engineering (6 years) and as general coordinator of the 12,600-megawatt Itaipu Project (6 years) and is now senior vice president and internal consultant. He has published more than 40 articles on dams and hydroprojects. He is a member of the National Academy of Engineering.

H. BOLTON SEED received his B.Sc. and Ph.D. at London University; he was awarded an S.M. at Harvard. Since 1950 he has been on the staff of the University of California at Berkeley; he is now professor of civil engineering. He has been a member of teaching staffs at London University and Harvard University. He also serves as a consultant to numerous major engineering companies and government agencies. He is the author of more than 180 papers and has been the recipient of 12 awards by the American Society of Civil Engineers for research contributions. He has been a member of the National Academy of Engineering since 1970.

ROBERT L. SMITH holds BSCE and MS (hydraulics) degrees from the University of Iowa. He has held several important positions as a consultant to various levels of government and in academia. He currently is Dean Ackers Professor of Civil Engineering at the University of Kansas. He is expert in applying civil engineering to water resources problems, especially flood management and is also considered expert in hydrology and hydraulic engineering. Mr. Smith is a member of the National Academy of Engineering, the Commission on Engineering and Technical Systems, and the Water Science and Technology Board.

JERY R. STEDINGER received his Ph.D. in engineering from Harvard University in 1977. At Harvard he was a member of the Environmental Systems Program. Since 1977, he has been at Cornell University, where he is associate professor. His research includes multireservoir systems analysis, groundwater management, and many topics in stochastic hydrology. During 1983-1984 he was on sabbatical leave with the USGS Water Resources Division in Reston, Virginia, working on special problems in statistical hydrology. He was recently selected by the National Science Foundation as one of 200 Presidential Young Investigators.

TECHNICAL CONSULTANT

HOMER B. WILLIS is a consulting engineer in private practice. He holds a B.S. degree in civil engineering from Ohio University. In more than 38 years as an employee of the U.S. Army Corps of Engineers, he was involved in many aspects of engineering for dams. In his last assignment with the Corps (1973-1979) he directed the technical engineering activities for the Civil Works Program for the development of water resources, including the nationwide program for inspection of nonfederal dams.

Index